Principles of
Digital Audio

Principles of Digital Audio

by

Ken C. Pohlmann

Howard W. Sams & Co.
A Division of Macmillan, Inc.
4300 West 62nd Street, Indianapolis, IN 46268 USA

International Standard Book Number: 0-672-22388-0
Library of Congress Catalog Card Number: 85-50747

Edited by: *C. Herbert Feltner*

Printed in the United States of America

Preface

A book on digital audio? Is this really necessary? In today's problematic world of cruise missiles, chemical dump sites, acid rain, and fast food croissants is a book on digital audio really important enough to be written? Moreover, in a world where technology sometimes appears to create two problems for every one it solves, if such a book tended to condone technology, would it really be desirable? Not to mention—why *this* author, at *this* time. . . .

In 1974, as a student at the University of Illinois, I worked in the Experimental Music Studios with a PDP-8 computer, something laughable by today's standards. But with that primitive technology a small yet vital experiment took place. I connected a digital-to-analog converter to the output port, wrote a program to compute the binary values of a sine wave computed over 256 increments, and output a new amplitude value of the sine computation every millisecond. From a 3 inch speaker laying on the teletype came a digitally generated sine wave—an acoustic event—direct from numbers. Everything which had always made so much sense conceptually and philosophically—mathematics and music—suddenly coalesced for me. I was hooked. But it was just the beginning for me and infant digital audio technology.

By 1976 I had designed several computer music systems, both all-digital and hybrid. One of the systems, utilizing the Texas Instruments 980A, a powerful 16 bit minicomputer, used digital oscillators and envelope generators to generate unprecedented sounds. My Master's thesis was a feeble attempt to put down on paper what promises those kluged circuits concealed.

During the next four years I worked on and off on the celebrated sound synthesizer, the Sal Mar Construction. This truly historic instrument performed electronic music in real time using analog audio output circuits and a marvelously complex digital system which generated the musical composition, under human guidance. But the Sal Mar weighed one-half ton and stood 8 feet high, Dr. Martirano and I doubted whether many consumers would be interested in such a device, or a book about it.

In 1980, soon to be Director of the Music Engineering program at the University of Miami in Coral Gables, I started a semester course in digital audio electronics. To my knowledge it was the first regular college course anywhere to specifically address the questions of sampling frequency, quantization, aliasing, modulation methods, error protection, audibility of digitization artifacts, etc. I gained valuable experience in the art of relating the complexities of digital audio in clear and understandable terms. There wasn't any book available for the course, I wrote my own lecture notes.

In 1982 I began technical writing, writing mainly about digital audio for professional and consumer audio magazines. Soon I was a monthly columnist for several magazines, forced by necessity to explore every aspect of digital audio to get my word quota each month. I polished my writing style and even picked up a few devoted readers. Simultaneously I was doing more and more digital recording in recording studios and concert halls. Digital audio was quickly revolutionizing professional and consumer audio technology. The time was drawing near.

In 1984 I brought home a Compact Disc player, hooked it up, and sat down to listen. As I enjoyed the fidelity of the compact consumer medium before me, I remembered my experiments with the PDP-8 ten years earlier. To have so closely witnessed such a dramatic advance in technology, which promises to bring such tremendous entertainment and information to so many people—I picked up a pencil and started writing.

A book on digital audio? Well, it's about time, I say. And quite a nifty book it is—everything fundamentally important to an understanding of the art of digital audio engineering. It's a book for everybody—the student can learn from it, and the experienced practitioner can sneak a copy into his or her briefcase and maybe avoid an early retirement. I have tried to include at least a mention of all of the important facts pertaining to digital audio, as well as some speculations on where it might be headed. But when it comes to a fast-changing topic, such as digital audio, why bother to write it all down? As an engineer and an educator I can think of many reasons. One of them, frankly, is that this is a profit venture. Mom, even you are going to have to pay for yours. Sorry, my motorcycle needs new pistons, not to mention tires and a clutch. Free copies only to editors of major magazines and reviewers who write favorable reviews. Someone has to pay off this manuscript's R & D.

One final point. This is a fundamentals book, thus it only goes so far. That suggests a need for intermediate, and advanced texts. . . . Of course, that depends on the public response to this little opus. I sincerely hope people like it. Ten years is a long time to prepare for a book.

KEN C. POHLMANN

Introduction

Psychologists and other professionals who deal with the foibles of human beings often speak of the learning curve, the experience it takes for a modification of a behavioral trait to take hold. In other words, they recognize that it takes people a little while to adapt to something new. Technologists and other professionals who deal with the foibles of machines often have no conception of how people react to change. They themselves are accustomed to innovation and usually expect rapid adaptation and acceptance. If a new product bombs, they are chagrined, and mystified.

It is important to recognize that a learning curve is inevitably undergone by people exposed to new technology. Change is not always easy and technology has the important added element of intimidation for many people. An experienced typist may at first be very reticent about switching to word processing. And unless that new product introduction and ensuing transition is properly handled, the typist might end up back at the typewriter, unwilling to adapt to a new technique. In a more intangible way, our transition from analog audio to digital audio recording involves the same pitfall. While entirely new recording techniques are required of audio professionals, and a microprocessor-based programmable Compact Disc player might be initially bewildering to home listeners, there is another obstacle common to both groups. Digital audio sounds different from analog audio recordings. While past changes in technology such as the conversion from vacuum tubes to solid-state, and from monaural to stereo have resulted in audible differences, the analog to digital transition is much more apparent. In the past, the only reproduction we have known has been achieved through analog technology; with digital technology the nature of the resulting fidelity is altered. The very definition of high fidelity has been altered with it.

Whether or not a veil has been lifted from the sound, or a gross misrepresentation has been perpetrated, is still a matter of debate in some circles. Some audiophiles, for example, whose practices are motivated by the principles of engineering, and sometimes by the principles of alchemy, insist that digital audio is a step backward in sound reproduction. Other critics have wholeheartedly embraced the new technology as yielding immediate results of higher fidelity, as well as the promise of even more extraordinary advances in the quality of recorded sound. The point of contention is an elusive one, namely, what it is we hear when we listen to recorded sound? Everybody agrees that the Compact Disc is marvelous in terms of its convenience and longevity compared to the LP, but those considerable advantages are overlooked when it comes to a discussion of listening. The point is that listening is an intimate kind of perception, and some of us will only grudgingly accommodate any change in what we hear while others quickly adapt to and enjoy the new sound.

In other words, the learning curve for the appreciation of recorded sound is experienced at widely varying rates.

If it is true that the medium is the message, then with digital audio the message has definitely changed. Digital audio does not sound like analog audio and it would be a terrible mistake (and a difficult engineering exercise) to modify the digital circuits to make the resulting sound mimic analog reproduction. Rather, we must carefully re-think our entire approach to the art and science of sound recording. When an orchestra is recorded, then played back, should it sound like a live orchestra, or like a recording? The answer is not obvious when it is considered that with the technology now available, such as digital audio technology, a recording of an orchestra could sound much *better* than a live performance. Is it our intent to document a symphonic performance, or to re-create it with a fuller fidelity and greater sense of perspective than that available to a concert-goer ticketed in one seat? And with musical events specifically created in the recording studio for which no live paradigm exists, what standard can be applied to our evaluation of the "fidelity" of its playback? In other words, we agree that fidelity is required, but faithful to what?

In short, there are many questions which the advent of sophisticated audio technology has brought to bear upon both audio professionals and consumers. This book is meant to directly attack the learning curve required of digital audio and shorten the time it takes to reconcile the impact of the new technology. To properly evaluate the immediate and potential effects of digital audio the technical nature of the method itself must be fully understood. Only in that way are the misconceptions de-toothed and enlightened judgements made possible. Eight chapters are included in this book, seven of them deal specifically with technical matters while one chapter is included for the perspective which I feel is critical in fully reconciling the changes engendered by digital audio.

Chapter 1 is a basics chapter. Traditionally, sound has been created and perceived as an analog acoustic event; our technical discussion begins with identification of the nomenclature of acoustics. Given the fundamental aspects of sound and hearing, both analog events, analog information is contrasted with digital information. Then the means of dealing with digital information, number systems, and binary arithmetic are considered.

Chapter 2 presents the fundamentals of digital audio. While analog recording has always used certain principles, digital recording brings with it its own principles which are relatively unfamiliar in the analog audio field. Discrete time sampling, amplitude quantization, signal-to-error ratio, aliasing, dither, and other topics have little precedent in analog audio. This chapter introduces each of these topics in an understandable way.

Chapter 3 begins our journey through a complete audio digitization system, using a hypothetical design of a classic pulse code modulation (PCM) system. The hardware design of the encoding side of the system is used as a step by step outline of the system. Each of the building blocks of the encoding scheme are explained. The anti-aliasing filter, dither circuit, analog-to-digital (A/D) converter, sample and hold circuit, multiplexer, record processing and modulation circuits are explained, as are the transformations undergone by the audio signal as it passes from its analog acoustic origins to binary data ready for storage.

Chapter 4 explains the output processing required to reproduce the previously encoded signal as an analog signal. Once again, the hardware design is used as a guide to trace the step by step transformations. The signal is followed from its incarnation as digital data as it is read from the storage medium and undergoes a series of processing steps including demodulation, error detection and correction, digital-to-analog (D/A) conversion, and output filtering. The fidelity of the output signal is compared to the input signal and the performance criteria for an audio digitization system are outlined. In conclusion, other digitization methods such as floating point and delta modulation are considered.

Chapter 5 outlines the various storage mediums presently available to store digitally encoded audio data. A discussion of longitudinal, perpendicular, and isotropic magnetic recording precedes descriptions of the most prominently used magnetic systems, stationary and rotating head tape recorders, and hard disk computer systems. Optical disc storage holds great promise for audio storage; read-only, write-once, and erasable optical systems are presented. The media chapter concludes with a look at two transmission media: direct satellite broadcast and the cable digital audio/data transmission system.

Chapter 6 is a software chapter; it features a discussion of error detection, correction, and concealment; data manipulations unprecedented in audio technology. The nature of errors are explained as are the basic techniques of error detection such as parity, interleaving, and redundancy. Error correction schemes such as block and convolutional codes are discussed in theoretical terms, as are practical methods such as Cross-Interleave Code. Error concealment techniques such as interpolation and muting conclude the chapter.

Chapter 7 is devoted to the Compact Disc, the consumer reproduction system designed to replace the LP as the dominant playback medium in the home, and extend its utility as a medium suitable for the automobile and portable playback. The nature of the data encoded on discs and the physical characteristics of discs is presented. Explanations of the theory of operation of the player are given, following the signal path of the data from the disc to the player's output with discussions of the laser pickup, EFM, CIRC error correction, and other topics.

Chapter 8 concludes the book with a discussion of the many questions, problems, and opportunities presented by the introduction of digital audio. Digital audio is more than a new technology, it is a fundamental change in our perception of recorded sound and a classic example of how new technology influences our expectations and perceptions. This chapter illustrates these important topics which will influence our acceptance of digital audio, as well as its future development.

Much of the material in this book stems from the work of the many pioneers and leaders in the field of digital audio technology. For their efforts in realizing the potential of this young science, we all owe a tremendous debt.

Contents

Chapter 3

Chapter 4

Chapter 1

Digital Audio Basics

Introduction

Digital audio is a highly sophisticated technology. It pushes the frontiers of engineering and manufacturing technique in integrated circuit fabrication, signal processing, and magnetic and optical storage. Although the underlying concepts have been firmly in place since the 1920s, commercialization of digital audio was postponed until the 1980s simply because theory had to wait 60 years for technology to catch up. Digital audio technology's complexity is all the more reason to start our discussion at the beginning. Although this book deals mainly with digital topics, we must include at least one analog topic—sound. Once the nature of sound is understood we can begin to explore ways to encode the information contained in an audio event and store it digitally. This leads us to the foundations of digital audio and topics such as binary numbers, sampling, quantization, aliasing, and dither.

1.1 Characteristics of Sound

It would be a mistake for a study of digital audio technology to forget the acoustic phenomena for which such a technology has been designed. Music is an acoustic event, whether it originates from instruments either radiating in air or creating electrical signals directly—all music ultimately finds its way into the air where it becomes a question of sound and hearing. It is appropriate to briefly review the fundamentals of the characteristics of sound to establish a common understanding of its nomenclature.

1.1-1 Sound Waves • Acoustics is concerned with the generation, propagation, and reception of sound waves. Sound may be thought of as a disturbance in a medium, in which energy is released to cause the creation of a pressure wave of alternating pressure above and below the equilibrium pressure. The molecules of the medium propagate the disturbance by their own successive displacements corresponding to the original disturbance; their displacement is in the direction in which the disturbance is travelling thus sound undergoes a longitudinal form of transmission, as shown in Fig. 1-1. A receptor placed in the sound field will similarly be moved according to the pressure impinging on it. Transducers, devices able to change energy from one form to another, may serve as sound generators and receivers. A violin changes the mechanical energy contributed by the bow to acoustical energy. A microphone may respond to the acoustical energy by producing electrical energy. A loudspeaker reverses that process to again create acoustical energy. Our ears accomplish the

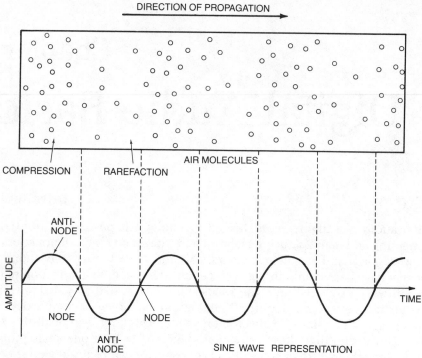

Fig 1-1. **Wave propagation.**

final transformation from acoustical energy to mechanical energy and ultimately to the electrical impulses sent to the brain, where the information in sound is perceived.

Sound may be produced either aperiodically or periodically. Thus, the outgoing pressure signal will either vary aperiodically or periodically. In either case, the velocity of sound in 68°F air is constant at 1130 feet per second. In the study of music, periodic waveforms are those most widely considered. One sequence of pressure compression and rarefaction of a periodic waveform determines one cycle which recurs. The number of recurrent cycles which pass a given point each second is the frequency of the sound wave, measured in Hertz (Hz). Alternatively, the reciprocal of frequency, the time it takes for one cycle to occur, is called the period. Given the velocity of a sound, and the frequency of a given sound, we may determine its wavelength, that is, the distance the sound travels to complete one cycle, or the physical measurement of the length of one cycle. Specifically, wavelength is the velocity of sound divided by the frequency of the sound, as shown in Fig. 1-2. Quick calculations demonstrate the enormity of the differences in the wavelength of sounds. A 20,000 Hz (20 kHz) signal is about one-half inch long while a 20 Hz signal is over 56 feet long. Few transducers are able to linearly produce or receive that range of wavelengths. Their frequency response is not flat, and the frequency range is limited. The range between the lowest and highest frequencies a system can accommodate, defines a system's bandwidth. To quantify that measurement, the deviation from flat response is specified according to application. Fig. 1-3 illustrates an audio device with flat response from 60 Hz to 9 kHz. However, its bandwidth might be specified as 20 Hz to 20 kHz.

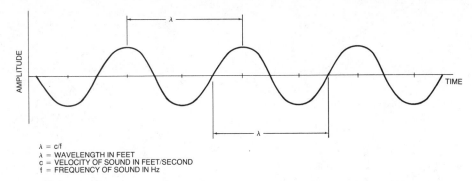

λ = c/f
λ = WAVELENGTH IN FEET
c = VELOCITY OF SOUND IN FEET/SECOND
f = FREQUENCY OF SOUND IN Hz

Fig. 1-2. **A periodic waveform characterized in terms of its frequency and wavelength, and the speed of sound.**

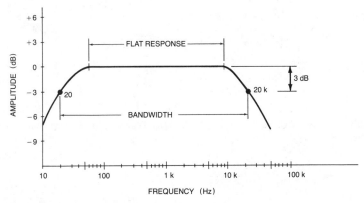

Fig. 1-3. **Bandwidth.**

Another important consideration of sound is its amplitude, that is, the amount of pressure displacement above and below the equilibrium level. Sound pressure at a point is the total instantaneous sound pressure minus the static (atmospheric) pressure. Sound pressure is very small because the amount of particle displacement in the medium is very small. In normal conversation, particle motion is only about one-millionth of an inch, while a crowd's acoustic outpouring might still be only about one-thousandth of an inch. Just as any system, such as our ears, has a frequency response, it has a dynamic range which measures the range of amplitude it can handle, from the softest to the loudest sound pressure levels.

1.1-2 Other Phenomena • Sound may be reflected and absorbed in a variety of ways. For example, the mere passage of sound through air acts to attenuate sound's energy. High frequencies are more prominently attenuated in air; a thunder clap heard close by as a sharp clap of sound whereas one far away is heard as a low rumble because the high frequencies have been attenuated by the air. In a free field, such as outdoors, sound continues to radiate outward from its source. The sound pressure level decreases with the inverse square of the distance. In an enclosure, such as a room, some energy is reflected back to create ambient information such as echoes and reverberation. A highly reflec-

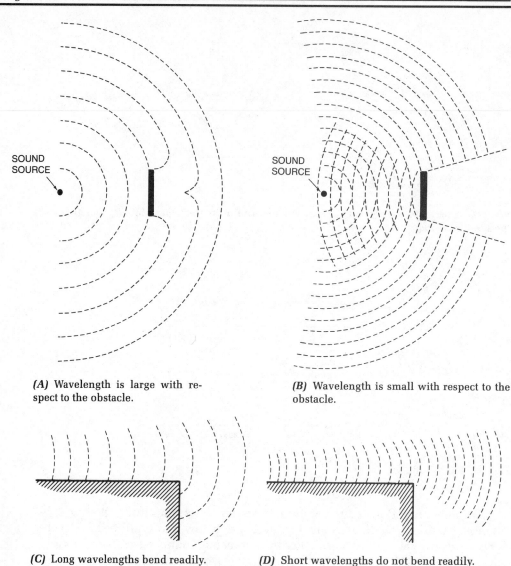

(A) Wavelength is large with respect to the obstacle.

(B) Wavelength is small with respect to the obstacle.

(C) Long wavelengths bend readily.

(D) Short wavelengths do not bend readily.

Fig 1-4. **Diffraction of sound.**

tive room with glass and tile creates conditions for a long reverberation time and multiple echoes. An absorptive room with carpet and curtains will have a short reverberation time, with few echoes.

Sound will undergo diffraction, in which it bends around obstacles or diffuses after passing through a small opening, as shown in Fig. 1-4. Diffraction is relative with respect to wavelength; longer wavelengths diffract more apparently than shorter ones. Thus, high frequencies are considered to be more directional in nature. Sound refracts, that is, changes direction with temperature changes because of the change in velocity it must undergo. Specifically, velocity of sound in air increases by about 1.1 feet per second with each increase in degree Fahrenheit. Because of the velocity change, sound will tend to bend away from warmer temperatures, and toward cooler ones, as shown in Fig. 1-5.

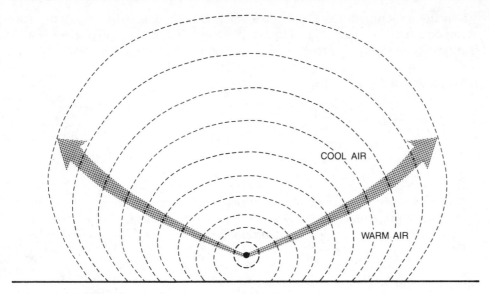

(A) Sound is refracted upward.

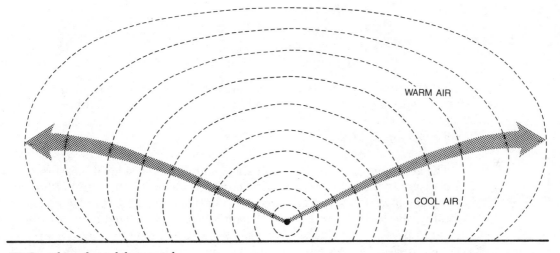

(B) Sound is refracted downward.

Fig. 1-5. **Refraction of sound.**

1.1-3 The Decibel • The characteristics of sound require a measuring unit able to accommodate the large range of values we encounter in electrical and acoustical systems. The decibel (dB) uses base 10 logarithmic units to achieve this. A base 10 logarithm is the power to which 10 must be raised to equal the value. For example, an unwieldy number such as 100,000,000 yields a logarithm of 8 ($10^8 = 100,000,000$). Specifically, the dB is defined to be one-tenth the logarithm of a power ratio, as demonstrated in the following:

$$\text{Level} = 10 \log \frac{P_1}{P_2} \text{ dB}$$

where P_1 and P_2 are values of acoustical or electrical power.

If the denominator of the ratio is set to a reference value, standard measurements may be made. For example, if $P_2 = 0.001$ watt is the reference, and a microphone's output is measured to be 0.0000001 watt, this is equivalent to:

$$\text{Power level} = 10 \log \frac{P_1}{P_2} \text{ dB}$$

$$= 10 \log \frac{10^{-7}}{10^{-3}}$$

$$= 10 \log 10^{-4}$$

$$= 10 \text{ X} -4$$

$$= -40 \text{ dB}$$

In acoustical measurements, intensity levels (IL) may be measured in dB by setting the reference intensity to 10^{-12} watts/m^2 (threshold of hearing) thus the intensity level of a rock band producing a sound of 10 watts/m^2 may be calculated:

$$\text{Intensity level} = 10 \log \frac{P_1}{P_2} \text{ dB}$$

$$= 10 \log \frac{10^1}{10^{-12}}$$

$$= 10 \log 10^{13}$$

$$= 10 \text{ X} 13$$

$$= 130 \text{ dB SPL}$$

When ratios of currents, voltages, or sound pressures are used, quantities whose square is proportional to power, the decibel formula becomes:

$$\text{Level} = 20 \log_{10} \frac{P_1}{P_2} \text{ dB}$$

The zero reference level for acoustical sound pressure level measurement is a pressure of 0.0002 dyne/cm^2. This corresponds to the threshold of human hearing, the lowest SPL we can perceive—it is equal to 0 dB SPL. The threshold of feeling, the loudest level before discomfort begins, is rated at 120 dB SPL. All sound pressure levels that we normally perceive may be rated on a scale, as shown in Fig. 1-6. We can characterize common acoustical environments in terms of their SPL: a quiet home might have an SPL of 35 dB, a busy street might be 70 dB SPL, and the sound of a jet engine in close proximity might exceed 160 dB SPL. An orchestra's pianissimo might be 30 dB SPL, but a fortissimo might be 110 dB SPL, thus its dynamic range is 80 dB.

The logarithmic nature of these decibels should be considered; they are not commonly recognizable, insofar as they are not linear measurements. If two motorcycle engines, each producing an IL of 80 dB, were started together, the combined IL would *not* be 160 dB. Rather, the logarithmic result would be a 3 dB increase, yielding a combined IL total of 83 dB. Of course, in terms of linear units, those two motorcycles each producing sound intensities of 0.0001 watt/m^2 would combine to produce 0.0002 watt/m^2. You are right. It's confusing.

1.1-4 Hearing • Our ear's response to frequency is logarithmic; this can be demonstrated through its perception of musical intervals. For example, the interval between 100 and 200 Hz is perceived as an octave, so is the interval between

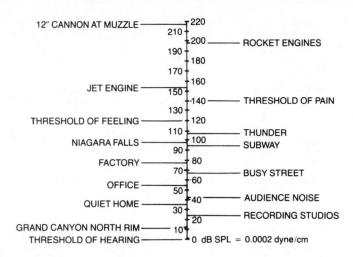

Fig. 1-6. Sound pressure levels—dB SPL.

1000 and 2000 Hz. In linear terms, the second octave is much larger, yet the ear hears it as the same interval. As we have seen, our ears can accommodate a very great amplitude dynamic range. For example, the threshold of feeling at 120 dB SPL is 1,000,000,000,000 times louder than the threshold of hearing at 0 dB SPL. In terms of convenience of expression, it is clear why we use the logarithmic decibel when dealing with that extreme range.

While our ears can be credited for their extended dynamic range and frequency response, their frequency response is, unfortunately, not very flat. In fact, they exhibit maximum sensitivity in the band from 3 to 4 kHz, and are relatively insensitive at low and high frequencies. In addition, the response varies with respect to loudness; the louder the sounds, the flatter our response becomes. Through testing, equal loudness contours, such as the Fletcher-Munson curves (also called equal-loudness contours), have been derived, illustrating these nonlinearities.

All characteristics of sound except one may be perceived by one ear; localization can only be accomplished with two ears. Using multiple cues such as intensity differences, waveform complexity, and time delays, the ear-brain interface may locate the placement of sounds in space. For example, a sound to the left will be louder in the left ear, and will be received earlier at the left ear; those cues are deciphered and the sound is judged to come from the left. Using these psychoacoustic principles, two loudspeakers may create phantom images, sound sources which appear to come from between the loudspeakers. Thus, stereo imaging may be created.

1.1-5 Phase • Two waveforms, identical in both shape and amplitude, may in fact be quite different because of their phase. If they have been delayed relative to each other in time, phase shift has occurred. Phase shift is often measured in degrees, as shown in Fig. 1-7. If two waveforms were combined, and their relative phase altered, a new waveform would result from constructive and destructive interference. Phase shift is thus the effect of relative time delay, with respect to two signals, either acoustically or electrically analog in nature. It can alter the nature of waveforms hence it can cause distortion. Absolute time delay holds no such danger; for example, the time delay between the

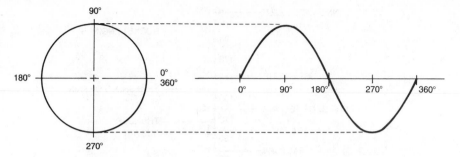

(A) One period is divided into 360 degrees, in both a circle and a periodic waveform.

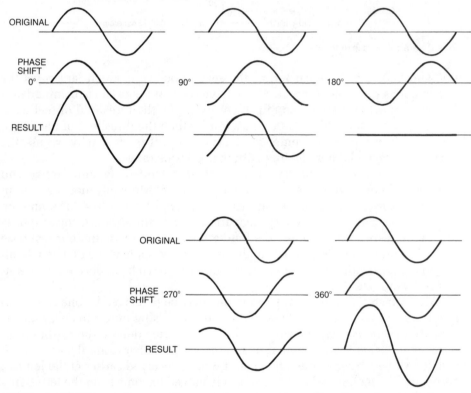

(B) Phase shift between two waveforms produces a new waveform through cancellation and reinforcement.

Fig. 1-7. **Characterization of phase shift.**

making of a studio recording and playback in the home is quite long, but irrelevant to the waveform's structure.

1.1-6 Complex Waveforms • The simplest form of periodic motion is the sine wave; it is exhibited by simplest oscillators, such as pendulums and tuning forks. The sine wave is remarkable because it exists only as a fundamental frequency. All other waveforms are complex and are comprised of a fundamental frequency and a series of harmonically related overtones. It is the relative amplitudes of overtones, and their phase relationships which uniquely create a waveform's timbre. For example, when the third harmonic is added to the fundamental, a complex waveform results. If the third harmonic is displaced in time (phase shift) and added to the fundamental, yet another complex waveform is created, as demonstrated in Fig. 1-8. The two complex waveforms shown would have the same pitch, yet sound dissimilar. Thus, a clarinet and trumpet may both play a note with the same fundamental; however, their sounds are obviously different because of their overtone series. This is summarized in the Fourier theorem which states that all complex periodic waveforms are comprised of a harmonic series of sine waves.

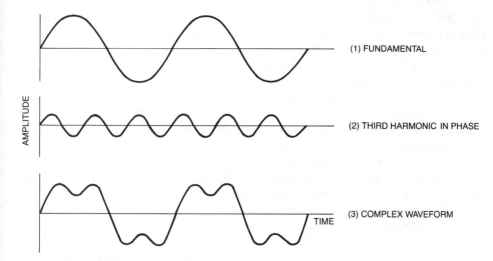

(A) When a fundamental sine wave is added to its third harmonic with one-third amplitude and no phase shift, the resultant complex waveform is produced.

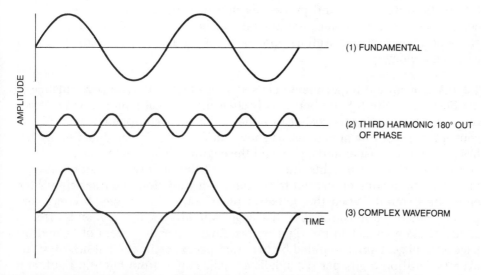

(B) When the third harmonic is shifted 180° a new complex waveform is produced.

Fig. 1-8. Complex waveforms are composed of harmonically related sine waves.

1.1-7 Sound Recording • A description of a hypothetical recording system might help to clarify some of the technique and terminology. In a multitrack session, multiple microphones are placed in the performing space to record a performance. Each microphone's signal passes through the console and is recorded on a multitrack tape recorder, one track for each microphone. Following the performance, the multitrack tape is replayed through the console and the signals are mixed to achieve proper balancing, a balance which might be better than any naturally achievable one. Overdubbing could be accomplished, in which new musical parts are added to the original music through synchronous monitoring of old tracks and recording of new tracks. The final mix is recorded onto a two track stereo tape recorder, and editing is then performed, to achieve the master recording.

The alternate technique, and the precursor to multitracking and overdubbing is direct to two track recording. For example, an orchestra might be stereophonically recorded with two or more microphones routed through a console directly to a two track tape recorder. In any case, the studio master tape is transferred to the appropriate production medium; for example, a tape for Compact Disc (CD) production is used to cut the CD master disc. The production master is delivered to the manufacturing facility where CDs, LPs, or tapes are mass produced and distributed to the consumer.

While the recording process appears to be quite simple, in practice it is an exacting and often frustrating endeavor. When a full fidelity recording reaches the hands of the consumer, it is because of the talented efforts of a few educated and experienced individuals. Digital audio technology holds the promise for much higher fidelity in recorded music; fidelity standards are thus more demanding with digital technology.

1.2 Analog Versus Digital

Digital audio entails entirely new concepts and techniques distinct from those utilized in analog audio technology. From the very onset, we must think of information, its storage and processing in a new light. In that respect, a few comparisons between analog and digital may be helpful in showing the differences between them, as well as some of the advantages and disadvantages of each technology.

1.2-1 A Conceptual Experiment • Let's suppose that I attend a performance of the Beethoven Ninth Symphony. As I settle into my seat I am surprised to see two recording engineers seated to either side; each equipped with a highly conceptual recording apparatus, as shown in Fig. 1-9. The engineer to my left has a long roll of paper and a pen, and the engineer to my right has a notebook and a pen. The house lights are dimmed and the performance starts. As the music plays, the one on my left traces out a wiggling line; the line mimics the orchestra's sound. When the orchestra plays softly the wiggles become very minute, when the music is loud the wiggles boldly swing back and forth. In contrast, the engineer to my right writes down a steady stream of 1s and 0s, page after page. I am astounded by his rapid pace; I estimate that he is writing over 1½ million digits per second. The music comes from left and right, and grows louder and softer, but it is difficult to see any correlation to his digits; perhaps they are in code.

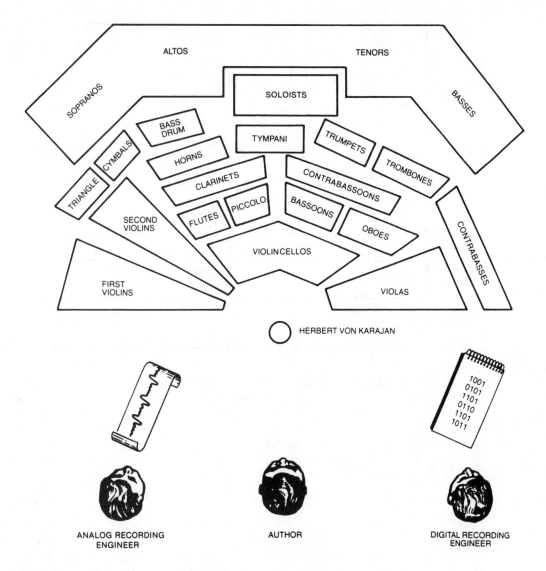

Fig. 1-9. **A conceptual experiment. Two methods of recording Beethoven's Ninth Symphony.**

With the final chords of the Ode to Joy the symphony ends, and both engineers lay down their pens as the house lights come on. Then both engineers return to the beginning of their documentation and prepare to play back their recordings. The engineer on my left lays the fingernail of his little finger on the wiggling line and begins to sing the Ninth. It is a good reproduction yet I notice that the paper's speed is uneven, and his fingernail cannot exactly trace the line. It is clear that no matter how careful he is, errors will always occur in both recording and reproduction because of the one-to-one nature of his method. The line can never be an exact physical replica of the acoustic event; therefore, it fails as a perfect means of documentation.

Meanwhile the other engineer has begun his playback. He flips through the notebook, page after page, merely reading the digits, singing an excellent

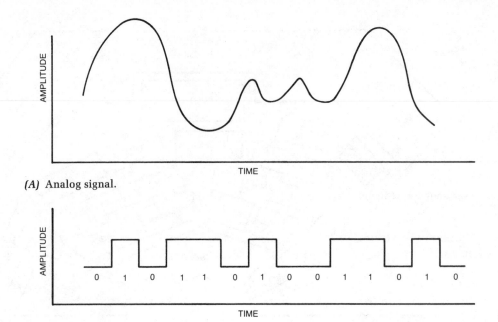

(A) Analog signal.

(B) Digital signal.

Fig. 1-10. **Analog and digital signals are two methods of representing information.**

Ninth. Furthermore, when he gets careless and varies the speed of his page turning the very precise rate of his read-out is unaffected. I notice that he blinks his eyes now and then and thus probably misses some digits, but I cannot perceive any audible error. Apparently, he is somehow correcting his mistakes. I recognize the benefit of his method: by converting to numbers he gains the advantage of the intelligibility and repeatability of a written quantified record over a replicated one. However, the task of recording all of those numbers is as monumental as the Ninth Symphony itself. With these observations my experiment ends.

Both engineers utilized methods to form a representation of the music. Both an analog (left method) and digital (right method) system aim for the same result of accurately storing and replaying an acoustical pressure waveform. However, their immediate goals are quite different. The analog system must strive to form a continuous replica of the acoustical waveform. The digital system periodically checks the waveform and records an approximation of the instantaneous value, thus its record differs considerably from the analog record, as shown in Fig. 1-10. The analog method provides one set of infinitely variable information whereas the digital method provides many discrete pieces of information. Compared to analog, the digital method is a mundane and unassuming way to record data, something that a bookkeeper might dream up. The method merely takes numbers and documents them for future reference. By the same token the digital method is highly efficient because the numbers with which it stores its information are easily saved and recalled.

1.2-2 An Analogy • An analog signal might be compared to a digital signal in terms of a bucket of water compared to a bucket of ball bearings, Blesser's analogy is shown in Fig. 1-11. Both water and ball bearings fill their respective

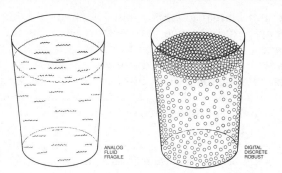

Fig. 1-11. A bucket of water and a bucket of ball bearings illustrate some of the differences between analog and digital information.

containers; the volumes of each bucket can be determined by the amount of their contents, but the procedures would be different. With water, we could weigh the bucket and water, pour out the water, weigh the bucket alone, subtract to find the weight of the water, then calculate what volume that represents. Or perhaps we could dip a measuring cup into the bucket and slowly withdraw measured amounts. In either case, we run the risk of spilling some water, or leaving some at the bottom of the bucket; our measurement would be imprecise.

With a bucket of ball bearings we could simply count each ball bearing, and calibrate the volume of the bucket in terms of the number of ball bearings it holds. The measurement would be relatively precise, if perhaps a little tedious (we might want to use a computer to do the counting for us). The ball bearings represent the discrete values in a digital system and point out the fact that with digital techniques we are able to quantify our values and gain more accurate information about our measurement. In general, precision is fundamental to any digital system. For example, a bucket which has been measured to hold 1.6 quarts of water is less useful than a bucket which is known to hold 8263 ball bearings. In addition, the bucket of ball bearings is a more permanent situation. If I tried to carry the water to another place I could easily spill some or even lose some to evaporation. The bucket of ball bearings could be more easily transported with less chance of mishap. The repeatability of each case is different, too. Once measured, I could attempt to re-create each situation. I could attempt to fill another bucket with water, but that bucket might be a different size; I could easily end up with 1.5 quarts. On the other hand, I could reliably count out 8263 ball bearings anywhere, anytime. We could continue the analogy, but the salient points should be clear. The utility of a digital system compared to an analog system is paradoxical: Conceptually, the digital system is much simpler because counting numbers is easier than dealing with a continuous flow, that is, it is easier to use data to represent the values of your task than to actually deal with the values themselves. But in practice, the equipment required to accomplish that simple task must be more sophisticated than any analog equipment.

Of course, audio recording and reproduction technology involves more than buckets. But the comparison between analog and digital audio may be simply summarized. An analog signal chain consisting of microphone, preamplifier, console, master tape recorder, master LP, pressed LP, phonograph player, amplifier, and loudspeaker, is an analog chain in which a continuous representation of amplitude change in time must be maintained. This is problematic because every circuit and storage medium throughout the chain is itself an analog device with its own analog distortion and noise which unavoidably contaminates the analog signal passing through. In short, when an analog signal is processed through a chain with inherent analog artifacts, degradation occurs.

With digital audio, the original analog event is converted into a stream of binary data which is then processed, stored, and distributed as a numerical representation. The reverse process, from data to analog event, occurs only at playback in the hands of the consumer thus eliminating intervening degradation. Since the analog information is carried through a numerical signal chain, it remains free of spurious analog distortion and noise—things not present in a numerical system. Of course, while digital systems are free of many analog problems, they can exhibit some anomalies of their own. For example, the representation of an infinitely variable amplitude with finite numbers must entail a degree of approximation.

1.3 The Binary Number System

Whereas acoustics and analog audio technology are mainly concerned with continuous mathematical functions to represent the waveform, digital audio is a study of discrete values. Specifically, a waveform's amplitude is represented as a series of numbers. That is an important first principle because numbers allow us to manage audio information very efficiently. Using digital computer techniques our capability to process this information has been greatly enhanced; the design nature of audio recording, signal processing, and reproducing hardware has followed the advance of digital technology and, for the first time, the idea of programming has been introduced into the practical audio environment. Thus, digital audio is foremost a numerical technology; to properly understand it we must first establish some groundwork with a discussion of number systems, focusing on the binary system used in digital computers.

1.3-1 The Meanings of Numbers • It all begins with numbers. When we deal with audio we are dealing with information, and numbers, as opposed to analog representation, offer a fabulous way to code, process, and decode information; in digital audio we use numbers to entirely represent audio information. We usually think of numbers as their symbols but the numerical symbols themselves are highly versatile; their meaning can vary according to the way we utilize them. For example, consider my classic BMW R50/2 motorcycle, built in 1962 with a 500 cc engine, and license plate 129907, as shown in Fig. 1-12. Obviously, there are a lot of numbers here, not so obvious is the important context of each of them. R50/2 represents the motorcycle's model number; 1962 is the year of manufacture; the number 500 represents the quantity of cubic centimeters of engine displacement; and the license number represents still another kind of information, a specifically coded information such that my speeding tickets are properly credited to my account. These various numbers are useful only by virtue of their arbitrarily assigned contexts. If that context is confused then information encoded by the numbers goes awry; I could end up with a motorcycle with license number 1962, manufactured in the year 500, with an engine of a cubic displacement of 129907 cubic centimeters. Similarly,

Fig. 1-12. A BMW R50/2 motorcycle.

the numerical operations we perform upon numbers is a question of interpretation; the tally of my moving violations determines when my license will be suspended, but the sum of my license plate numerals is probably harmless. Numbers, if properly defined, provide a good method to store and process data. The negative implication is that the numbers and their meanings have to be used carefully.

1.3-2 Number Systems • For most of us, the most familiar numbers are those of the base 10 system, perfected in the Ninth Century by some clever Arabs who suddenly conceived of the "0" numeral to represent nothing and appended it to the nine other numerals already in use. Early societies were stuck with the unitary system which used one symbol, in a series of marks, to answer the essential question—how many? That is an unwieldy system for large numbers thus higher base systems were devised. The Mesopotamians, who considered themselves to be a fairly advanced bunch, invented a number system which used 60 symbols. It was a little cumbersome, but even today, 3700 years later, we still use the essence of their system to divide an hour into 60 minutes, a minute into 60 seconds, and a circle into 360 degrees.

A number system is essentially a question of preference, because any integer may be expressed using any base. The point is that choosing any number system involves the question of how many different symbols you think is most convenient. Our base 10 system uses 10 numerals; we say that the radix of the system is 10. In addition, the system uses positional notation; the position of the numerals tells us the quantities of ones, tens, hundreds, thousands, etc. In other words, each next place is multiplied by the higher power of the base. A base 10 system is convenient for 10 fingered organisms such as humans, but other number bases may be more appropriate for other applications. Of course, you have to know the radix; the numerals "10" in base 10 is the total number of fingers we have, but "10" in base eight is the number of fingers minus the thumbs. Similarly, would you rather have 10,000 dollars in base 6, or 100 in base 60? Table 1-1 compares four number systems.

Table 1-1. Four Common Number Systems

Hexadecimal (base 16)	Decimal (base 10)	Octal (base 8)	Binary (base 2)
0	0	0	0000
1	1	1	0001
2	2	2	0010
3	3	3	0011
4	4	4	0100
5	5	5	0101
6	6	6	0110
7	7	7	0111
8	8	10	1000
9	9	11	1001
A	10	12	1010
B	11	13	1011
C	12	14	1100
D	13	15	1101
E	14	16	1110
F	15	17	1111

1.3-3 Base Two • Gottfried Wilhelm von Leibnitz, the great philosopher and mathematician, stumbled onto the binary number system on March 15, 1679; that day marks the origin of today's digital systems. While base 10 is handy for humans, a base 2, or binary, system is more efficient for digital computers, and digital audio equipment. Only two numerals are required, and they may efficiently satisfy the machine's principle electrical concern of *voltage/no voltage*, or *on/off*. A binary system is ruthlessly efficient for a machine, and fast. Imagine how quickly you can turn a switch on and off, that represents the rate you can process information. Or watch a square wave go by, which means a machine is operating the switch for you. And consider the advantages in terms of storage; instead of saving infinitely different analog values, you only have to remember two values. As we will see, the efficiency of binary data enables digital circuits to handle the tremendous amount of information contained in an audio signal.

Whatever information is being processed, in our case whatever kind of audio signal has been converted to binary data, no matter how unrelated it might appear to be to numbers, a digital processor, or computer, codes the information in the form of a number, using the base two system. To better understand how our audio data is being handled inside the digital audio system, a brief look at the arithmetic of base two will be useful. In fact, we will consistently see that the challenge of coding audio information in digital form is a central issue in the design and operation of digital audio systems.

In essence, all number systems perform the same function; thus, we may familiarize ourselves with the binary system by comparing it to our decimal system. A given number may be expressed in terms of either base system, and conversion from one base to another can be easily accomplished. Several methods may be used, an easy decimal to binary conversion for whole numbers is division of the decimal number by two, and collection of the remainders to form the binary number, as demonstrated in Fig. 1-13. Similarly, conversion from binary to decimal can be accomplished by writing the expression for the binary number in power of 2 notation, then expanding and collecting terms to form the decimal number, as demonstrated in Fig. 1-13B.

$$11010.001_2 = ?_{10}$$

$$11010 = 1 \times 2^4 + 1 \times 2^3 + 0 \times 2^2 + 1 \times 2^1 + 0 \times 2^0$$
$$= 1 \times 16 + 1 \times 8 + 0 \times 4 + 1 \times 2 + 0 \times 1$$
$$= 16 + 8 + 2$$
$$= 26_{10}$$

integer part

$$11010_2 = 26_{10}$$
$$.001 = 0 \times 2^{-1} + 0 \times 2^{-2} + 1 \times 2^{-3}$$
$$= 0 \times 0.5 + 0 \times 0.25 + 1 \times 0.125$$
$$= .125_{10}$$

fractional part

$$.001_2 = .125_{10}$$

thus $\quad 11010.001_2 = 26.125_{10}$

(A) Conversion of a decimal number to binary.

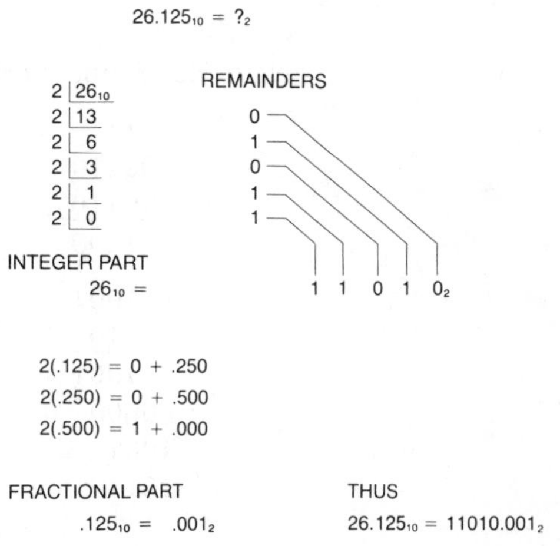

$26.125_{10} = ?_2$

	REMAINDERS
2 ⌊26₁₀	
2 ⌊13	0
2 ⌊ 6	1
2 ⌊ 3	0
2 ⌊ 1	1
2 ⌊ 0	1

INTEGER PART
$26_{10} =$ 1 1 0 1 0₂

2(.125) = 0 + .250
2(.250) = 0 + .500
2(.500) = 1 + .000

FRACTIONAL PART
$.125_{10} = .001_2$

THUS
$26.125_{10} = 11010.001_2$

(B) Conversion of a binary number to decimal.

Fig. 1-13. **Base ten/base two conversion.**

This points up the fact that the base two system also employs positional notation. In base two each next place represents a doubling of value. The right-most column represents 1s, the next column is 2s, then 4s, 8s, 16s, etc. Once again, it is important to designate the base being used; for example, in base two the symbol "10" could represent the total number of hands we have.

Just as a decimal point is used to delineate the whole number from the fractional number, a binary point does the same for binary numbers. Conversion of the fractional part of a decimal number to a binary number is done by multiplying the decimal number by two. Conversion often leads to an infinitely sustaining binary number, so we must limit the number of terms.

As in our base 10 system, the standard arithmetic operations of addition, subtraction, multiplication, and division are applicable to the base two system, as shown here:

0	0	1	1	
+0	+1	+0	+1	addition
0	1	1	10	

carry

borrow

1	1	0	10	
−1	−0	−0	−1	subtraction
0	1	0	01	

```
    0          0          1          1
   ×0         ×1         ×0         ×1      multiplication
    0          0          0          1


    0          1
   ÷1         ÷1                             division
    0          1
```

example of binary arithmetic

```
      11
   00110                              11011
  +10110                             −01101
   11100                              01110


    1001                              1011
   ×101                       101|110111
    1001                              101
   0000                              0011
   1001                              0000
  101101                              111
                                      101
                                     0101
                                     0101
                                        0
```

Just as in any base, the fundamental operation, addition, is easily carried out in base two because we have memorized addition rules to form an addition table. In base two, we are merely careful to utilize the addition rules unique to that base. The procedure is the same as in the decimal system, except it's easier because the addition table is simpler. There are only 4 combinations compared to the more than 100 possible combinations resulting from the rules of decimal addition. The generation of the carry, as in the decimal system, is necessary because the result in that case is larger than the largest digit in the system. The algorithms for subtraction, multiplication, and division in the binary system are identical to the corresponding algorithms in the decimal system.

Thus, a number is what we make it and the various systems differing only by their base operate in about the same way. A computer's use of the binary system is merely a question of expediency; it presents no real barrier to our understanding of digital techniques. It is simply the most logical approach. Ask yourself, would you rather deal with 10 voltage levels, 60, an infinitely analog number of them, or 2?

1.3-4 Boolean Algebra • The binary number system presents tremendous opportunities for the design of electronic equipment. Boolean algebra is the method used to combine and manipulate binary signals. It is named in honor of its inventor George Boole who published his proposal for the system in 1854 in a very curious work entitled: An Investigation of the Laws of Thought. The set of basic logic operations illustrated in Fig. 1–14 is defined using one or two variables. They may be used singly or in combinational logic to perform any possible logical operation. The *on/off* nature of the system is ideally suited for realization in electrical hardware. With the set of fundamental logic operations

COMPLEMENT $F = \bar{X}$

X	$\bar{X}$
0	1
1	0

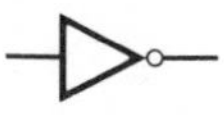

AND $\quad F = X \cdot Y$

X	Y	$X \cdot Y$
0	0	0
0	1	0
1	0	0
1	1	1

OR $\quad F = X + Y$

X	Y	$X + Y$
0	0	0
0	1	1
1	0	1
1	1	1

EXCLUSIVE OR

$F = X \oplus Y$

X	Y	$X \oplus Y$
0	0	0
0	1	1
1	0	1
1	1	0

NAND $\quad F = \overline{X \cdot Y}$

X	Y	$\overline{X \cdot Y}$
0	0	1
0	1	1
1	0	1
1	1	0

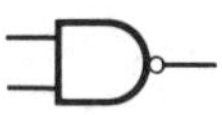

NOR $\quad F = \overline{X + Y}$

X	Y	$\overline{X + Y}$
0	0	1
0	1	0
1	0	0
1	1	0

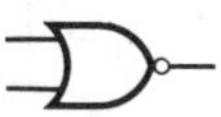

Fig. 1-14. **Boolean operators.**

to manipulate the BInary digiTS, or bits, we have the tools necessary to design the logic networks which comprise useful digital systems. Everything from hard-wired logic gates to microprocessors and supercomputers may be designed to take advantage of this efficient system.

Given a set of operators, our next step is to develop a system of algebraic relations which forms the basis of digital processing in the same way that regular algebra governs the manipulation of our familiar base 10 operations. In fact, the two systems are very similar. Relations such as complementation, commutation, association, and distribution as shown in Fig. 1-15 form the system of mathematical logic needed to create logical circuits and software. Given these fundamental abilities, the hardware of digital systems may be designed as shown in Fig. 1-16, and software written, to manage virtually any kind of information and solve any numerical problem. In another simple example, an adding circuit may be designed using NAND gates. Given two inputs X and Y and carry input C, the sum S and carry output C results, as shown in Fig. 1-17. Although computer design is beyond the scope of this book, we have

1. Special properties of 0 and 1

$$0 + X = X \qquad 0 \cdot X = 0$$
$$1 + X = 1 \qquad 1 \cdot X = X$$

2. Idempotence laws

$$X + X = X \qquad X \cdot X = X$$

3. Involution

$$\overline{\overline{X}} = X$$

4. Complementation laws

$$X + \overline{X} = 1 \qquad X \cdot \overline{X} = 0$$

5. Commutative laws

$$X + Y = Y + X \qquad X \bullet Y = Y \bullet X$$

6. Associative laws

$$X + (Y + Z) = (X + Y) + Z = X + Y + Z$$
$$X \cdot (Y \cdot Z) = (X \cdot Y) \cdot Z = X \cdot Y \cdot Z$$

7. Distributive laws

$$X \cdot (Y + Z) = (X \cdot Y) + (X \cdot Z) \qquad X + (Y \cdot Z) = (X + Y) \cdot (X + Z)$$

8. Absorption laws

$$X + (X \cdot Y) = X \qquad\qquad X \cdot (X + Y) = X$$
$$X + (\overline{X} \cdot Y) = X + Y \qquad X \cdot (\overline{X} + Y) = X \cdot Y$$

Fig. 1-15. **Laws of Boolean algebra.**

ORIGINAL EXPRESSION:

$$F = \overline{X} \cdot \overline{Y} \cdot \overline{Z} + \overline{X} \cdot \overline{Y} \cdot Z + X \cdot \overline{Y} \cdot Z$$

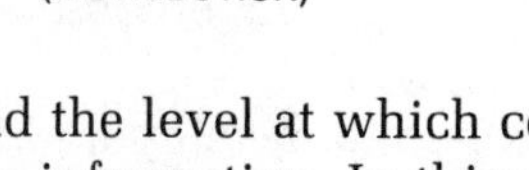

Fig. 1-16. **Complex logical expressions may be simplified with Boolean laws for simple hardware realization.**

MAY BE SIMPLIFIED:

$F = \overline{X} \cdot \overline{Y} \cdot \overline{Z} + \overline{X} \cdot \overline{Y} \cdot Z + \overline{X} \cdot \overline{Y} \cdot Z + X \cdot \overline{Y} \cdot Z$	(IDEMPOTENCE)
$F = \overline{X} \cdot \overline{Y} \cdot (\overline{Z} + Z) + (\overline{X} + X) \cdot \overline{Y} \cdot Z$	(DISTRIBUTION)
$F = \overline{X} \cdot \overline{Y} + \overline{Y} \cdot Z$	(COMPLEMENTATION)
$F = (\overline{X} + Z) \cdot \overline{Y}$	(DISTRIBUTION)

already seen enough to understand the level at which computers operate and the techniques they use to process information. In this respect all digital systems are identical. They differ, however, according to the kind of information they process and what data manipulations they are called upon to accomplish. A digital audio system is thus a unique digital system especially configured to manage audio data.

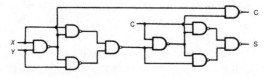

Fig. 1-17. **An adding circuit using** NAND **gates.**

Chapter 2

Fundamentals of Digital Audio

Introduction

The use of digital methods for the recording, reproduction, and storage of digital audio signals entails several concepts foreign to analog audio methods. In fact, digital audio systems bear little resemblance to analog systems, especially in terms of the critical function of each—the processing of audio information. Since audio itself is analog in nature, digital systems employ sampling and quantization, the twin pillars of audio digitization, to transform the audio information. Special precautions must be taken to combat two fundamental types of distortion, a condition of erroneous frequencies known as aliasing, and the error resulting from the quantization of the analog waveform.

2.1 Discrete Time Sampling

With analog recording we continuously modulate tape or cut a groove, but with digital we must use numbers. The first question we are faced with is how to choose numbers. In other words, how do we go about recording a data point from a changing waveform? Digitization employs time sampling and amplitude quantization to encode the infinitely variable analog waveform as discrete values in time and amplitude. We will consider these techniques in the following sections. First, let's consider the idea of sampling, the essence of digital audio.

2.1-1 The Lossless Nature of Sampling • Let's use a clock analogy to illustrate how sampling differentiates a digital music system from an analog system. Time seems to flow continuously. The hands of an analog clock sweep across the clock face covering all time as it passes by. A digital read-out clock also tells time, but with a discretely valued display. In other words, it displays sampled time. It is the same with music. Music varies continuously in time and may be recorded and reproduced either in continuous analog form or time-sampled digital form. Just as both clocks tell the same time, both types of recordings play the same music. Time sampling is the essential mechanism which defines a digital audio system, permits its analog-to-digital conversion, and differentiates it from an analog system.

A nagging question presents itself at this point. If a digital system samples discretely, what happens between samples? Haven't we lost the information occurring between sample times? The answer, intuitively surprising, is no—

given correct conditions, no information is lost due to sampling between the input and output of a digitization system. The samples contain the same amount of information as the conditioned unsampled signal. To illustrate this, let's try another conceptual experiment.

Let's suppose that we fasten a movie camera on the handlebars of our BMW motorcycle and go for a drive, up and down hill, over smooth pavement and some not so smooth, then return home and process the film. When we audition our piece of avant-garde cinema we discover that the discrete frames of film successfully merge to reproduce our ride; it looks great. But when we come to some bumpy pavement, our picture is blurred, we ascertain that the quick movements were too fast for each frame to capture the change. We draw the following conclusions: if we increase the film speed, using more frames per second, we would be able to capture quicker changes. If we complained to City Hall and had the bad pavement smoothed, then there would be no blur even at slower frame speeds. Our movie would perfectly reproduce our motorcycle ride. We settle on a compromise—we make the roads reasonably smooth, then use a frame rate adjusted for a clean picture.

Just as the discrete frames of a movie create a moving picture, the samples of a digital audio recording create time varying music; there is little conceptual difference between the visual and aural systems. Just as no information is lost between the frames of a properly shot motion picture, nothing is lost between the samples of a digital audio recording. As we discussed, sampling is a lossless process if the signal is properly conditioned. Thus, in a digital audio system we must smooth out the bumps in the incoming signal, specifically it is low-pass filtered. That is, the frequencies too high to be properly sampled are removed. We design the system so that the threshold of these filtered frequencies is above the limit of human audibility.

2.1-2 The Sampling Theorem • When the input signal is low-pass filtered, we can theoretically sample the signal such that there is no loss of information due to sampling between the output sampled signal and the input smoothed signal. From a sampling standpoint, it is not an approximation; it is exact as stated by the Nyquist sampling theorem. The discrete time method of sampling defines only instantaneous values. However, it can be mathematically proven that a sampled band limited signal contains the same amount of information as the original unsampled smoothed signal. When the signal is smoothed, we can compute all the intervening values without error and thus re-create the original waveform. Consider the waveform in Fig. 2-1. The continually changing analog function has been sampled to create a series of pulses; the amplitude of each pulse when chosen from a vertical scale will ultimately yield a number which represents the analog amplitude at that instant. To quantify the situation, we define the sampling rate as the number of samples per second. Its reciprocal, sampling time, is the time between each sample. For example, a sampling time of $1/40,000$ second corresponds to a rate of 40,000 samples per second. It is apparent that a quickly changing waveform, that is, one with high frequencies, would require a shorter sampling time, as we saw in our motorcycle movie. Thus, sampling rate determines the frequency response, and overall signal throughput bandwidth of the digitization system. The choice of sampling rate is one of the most important design criteria of a digitization system since it determines bandwidth of the system. Thus, the question is presented—how often should we sample to accurately represent a music waveform?

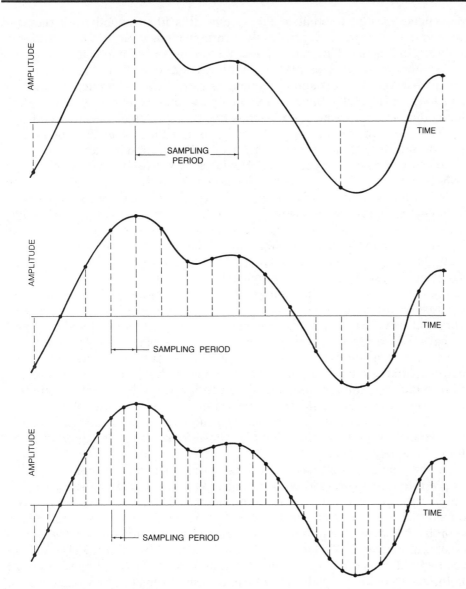

Fig. 2-1. **Discrete time sampling.**

Sampling theory answers the question of sample rate; Nyquist has shown that S samples per second are needed to completely represent a waveform with a bandwidth of S/2 Hz. In other words, we must sample at a rate twice the highest throughput frequency to achieve lossless sampling. Thus, an audio signal with a frequency response of 0 to 20 kHz would theoretically require a sampling rate of 40 kHz for proper digital encoding. It is crucial to observe the sampling theorem's criteria for limiting the input signal to no more than half the sampling frequency. Just as a bumpy road blurred our movie, too high a frequency in a digitization system would cause distortion. This is examined in greater detail in Section 2.2. A low-pass filter always precedes the sampling circuit to remove frequencies above the half-sampling frequency limit. A low-pass filter is also placed at the output of a digital audio system to remove high frequencies created internal to the system; this filter smooths the staircase effect in the reconstructed sampled waveform to recover the original waveform, as shown in Fig. 2-2. This is discussed in more detail in Chapter 4.

Another question presents itself with respect to the sampling theorem. It can be observed that low audio frequencies can be easily sampled; because of their long wavelengths, there will be many samples to represent each period. But as the sampled frequencies become higher, the periods are shorter. There will be fewer samples per period. Finally, in the theoretical limiting case of critical sampling, at an audio frequency of half the sampling frequency, there will be only two samples per period. However, even two samples can represent a waveform. For example, consider the case of a 40 kHz sampling system and an audio sine wave input at 20 kHz, as shown in Fig. 2-3. The digitizer would produce two samples which would be used to construct a 20 kHz square wave. In itself, this reconstructed waveform is quite unlike the original sine wave. However, the 20 kHz square wave is comprised of odd harmonics, sine waves at 20, 60, 100, 140, and 180 kHz, etc. A low-pass filter at the output of the digital audio system removes all frequencies higher than those originally entering the system. All higher harmonics are gone. The output of the system is a 20 kHz sine wave, the same waveform as originally input. We know that the 20 kHz input waveform was a sine wave because the input low-pass filter would not have passed higher waveform harmonics to the sampler. As far as our ears are concerned, a sine wave is perfectly suitable because the harmonics of any complex 20 kHz waveform are above our range of audibility anyway. Even in the limiting case, the sampling theory is valid. Thus, it is correct to state that higher sampling rates would permit recording and reproduction of higher frequencies. But given the design criteria of an audio frequency bandwidth, higher rates would not improve the fidelity of those frequencies already within the allowed frequency range.

In the previous example of critical sampling, there is no guarantee that the sample times will coincide with the maxima and minima of the waveform. Samples could come from lower amplitude parts of the waveform, or even coincide with the zero axis crossings of the waveform. In practice this poses no problem. Critical sampling is not attempted; a sampling margin is always present. Moreover, as we shall see, settling time of the sampling circuits, the time required for proper signal manipulation, provides a sampling window 10 microseconds or longer, yielding sufficient margin to obtain a sample from even the 25 microsecond half period of a 20 kHz waveform.

As we have seen, to satisfy the sampling theorem, manufacturers must design a low-pass filter into any digitization system, placing it first in the signal chain. Because these analog filters cannot cut off the signal as suddenly as the sampling theorem demands, a frequency guard band is employed. The filter's cutoff frequency characteristic is begun at a lower frequency allowing several thousand Hz for the filter to sufficiently attenuate the signal to make certain that no frequency higher than half the sampling rate enters the digitization circuitry.

It should be emphasized that the need to low-pass filter the audio signal is not as detrimental as it might first appear. As long as we choose an appropriate sampling rate, we can extend the frequency response of the audio signal as far as we wish. The trade-off, of course, is the demand we place on the digital circuitry and the storage medium. Higher sampling rates require that circuitry

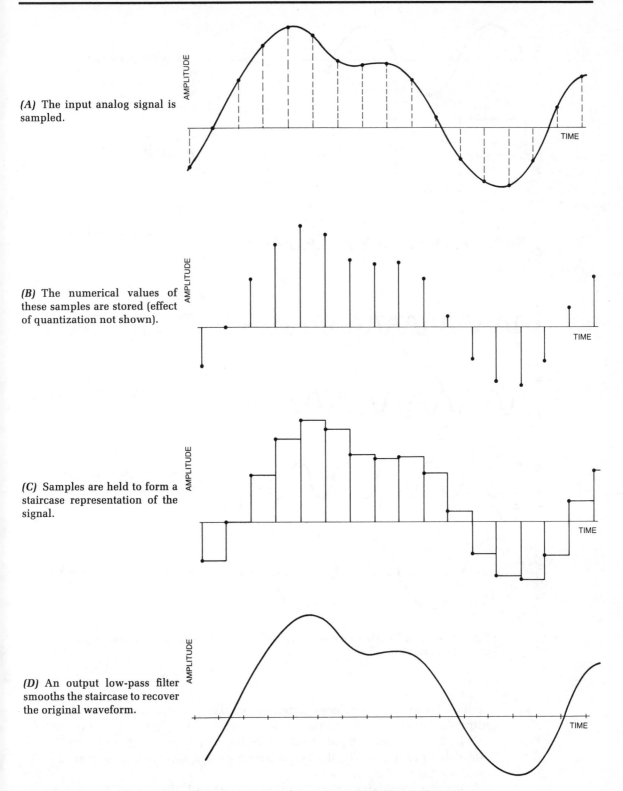

(A) The input analog signal is sampled.

(B) The numerical values of these samples are stored (effect of quantization not shown).

(C) Samples are held to form a staircase representation of the signal.

(D) An output low-pass filter smooths the staircase to recover the original waveform.

Fig. 2-2. Discrete time sampling: a band limited signal can be sampled and reconstructed without loss.

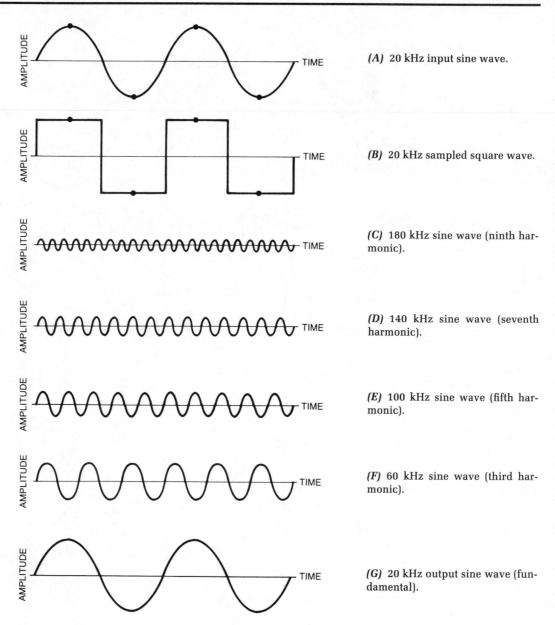

(A) 20 kHz input sine wave.

(B) 20 kHz sampled square wave.

(C) 180 kHz sine wave (ninth harmonic).

(D) 140 kHz sine wave (seventh harmonic).

(E) 100 kHz sine wave (fifth harmonic).

(F) 60 kHz sine wave (third harmonic).

(G) 20 kHz output sine wave (fundamental).

Fig. 2-3. **When a 20 kHz sine wave is sampled at 40 kHz rate, the two sample points reconstruct a square wave.**

operate faster and that larger amounts of data be stored; both of these are ultimately questions of economics. Manufacturers have chosen a sampling rate of 44.1 kHz for the Compact Disc, for example, because such a system can be affordably produced. The flat audio frequency response of 0 to 20 kHz is sufficient for most listeners.

Fig. 2-4 summarizes the sampling process and illustrates a further ramification: sampling creates duplicates of the input waveform's spectrum (frequency response curve) at multiples of the sampling frequency (Fig. 2-4F). These high frequency images are removed with another low-pass (anti-imaging) filter at the output of the digitization system.

At any rate, the point is clear, a band limited signal may be sampled, stored as discrete values, de-sampled, and reproduced. No information is lost through sampling. Sampling theorems, such as the Nyquist Theorem, demonstrate this conclusively. Of course, time sampling is only half the battle. A digital system must also determine the actual numerical values it will use at sample time to

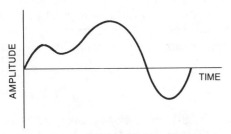

(A) Input waveform (after anti-aliasing filter).

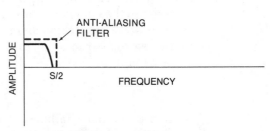

(B) Spectrum of input waveform.

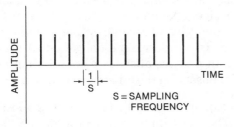

(C) Sampling signal.

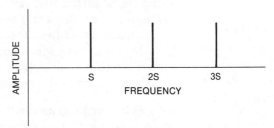

(D) Spectrum of sampling signal.

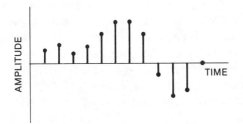

(E) Sampled input waveform.

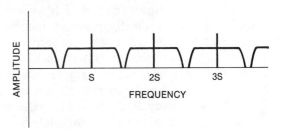

(F) Spectrum of sampled input waveform.

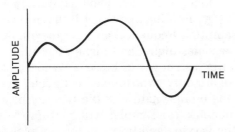

(G) Output waveform (after anti-imaging filter).

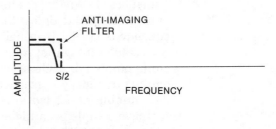

(H) Spectrum of output waveform.

Fig. 2-4. **Time and frequency domain signals of sampling.**

represent the original waveform's amplitude. This question of quantization is explained later in this chapter.

2.1-3 Fast Sampling Rates • Before we "close the book" on discrete time theory, we should mention a current hypothesis concerning the nature of time. We mentioned that time *seems* to be continuous. However, some physicists have recently suggested that like energy and matter, time might come in discrete packets. Just as this book consists of a finite number of atoms, and could be converted into a finite amount of energy, the time it takes you to read the book might consist of a finite number of time particles. Specifically, the indivisible period of time might be 1×10^{-42} second (that's a 1 preceded by a decimal point and 41 zeros). The theory is that no time interval can be shorter than that because the energy required to make the division would be so great that a black hole would be created and the event would be swallowed up. If any of you out there are experimenting in your basements with *very* fast sampling rates, please be careful.

2.2 Aliasing

One particular challenge to the audio digitization system designer is that of aliasing, a kind of sampling confusion that can take place in the recording side of the signal chain. Just as a criminal can take many names, and thus confuse his identity, aliasing can create false signal components. These erroneous signals can appear within the audio bandwidth and are impossible to distinguish from legitimate signals. Obviously, it is the designer's obligation to prevent such distortion from ever occurring. In practice aliasing is not a limitation. It merely underscores the importance of observing the criteria of sampling theory.

2.2-1 Foldover Frequencies • We have demonstrated that sampling is a lossless process, under certain conditions. The most important condition is that the input signal must be band limited, that is, a low-pass filter must precede the sampling circuit. If this is not done, some highly undesirable effects can result. Specifically, aliasing distortion can occur. High frequency changes in amplitude will not be properly encoded thus information will be lost. However, lost information is the least of our worries. The under sampling also creates entirely new erroneous signals, another form of distortion. Let's consider another conceptual experiment: you take a movie of me while I get on my BMW and drive away. In the film, as I accelerate, the wheels rotate forward, appear to slow and stop, then begin to rotate backward. . . .

Aliasing is a consequence of a disallowed condition in sampling theory; Nyquist has shown that the highest throughput frequency in a sampling system can only be equal to or less than half the sampling frequency. If the throughput frequency is greater than half the sampling frequency, aliasing inevitably occurs. As the throughput frequency becomes higher and higher, the number of sample points per cycle becomes fewer, as shown in Fig. 2-5. When half the sampling frequency is reached, there are only two samples per cycle, the absolute minimum needed to record the bipolar nature of the waveform. If we attempt to sample even higher frequencies, the sampler will continue to produce samples at its fixed rate, but the varying amplitude of the samples from those deviant audio frequencies creates false information for new frequencies. As the deviant frequencies go higher and higher, new descending frequencies

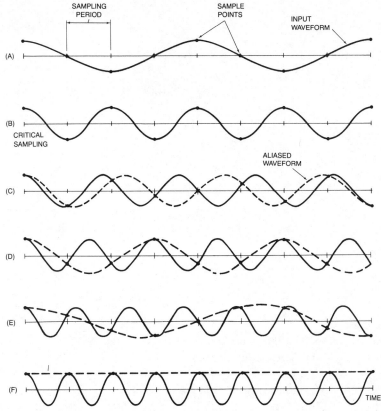

Fig. 2-5. Aliasing. If frequencies greater than the half sampling frequency are sampled, aliased frequencies (dotted curves) are created.

are created. Specifically, if S is the sampling rate, and F is a frequency higher than half the sampling rate, then a new sampled frequency F_a is also created at $F_a = S - F$. For example, If S = 44 kHz, and we attempt to sample a 36 kHz signal, another sampled frequency appears at 8 kHz. If we attempt to sample a 40 kHz signal, 4 kHz appears. In other words, a new aliased frequency appears back in the audio band, folded over from the sampling frequency. In fact, it is sometimes called foldover.

To elaborate upon this scenario, in which false frequency information accompanies the actual frequency following the sampler, we must remember that a low-pass filter is used at the output of a digitization system to smooth the staircase function and thus recover the original signal. That output filter will be designed to cut the sampling frequency in half. Thus, errant input frequencies above that value will be filtered out. We are left only with the aliased frequency in the audio band. And that isn't all—but before we consider further ramifications, let's take a look at an example.

Suppose we have a digitization system sampling at 44 kHz. Further, suppose that a signal with a frequency of 36 kHz has somehow sneaked into our sampler as shown in Fig. 2-6. Our sampler would produce the improper samples, faithfully recording a series of amplitude values at sample times. Given those samples, no device, digital or otherwise, could decide which was the intended frequency—36 kHz, or 8 kHz? After the output filter, the 36 kHz signal

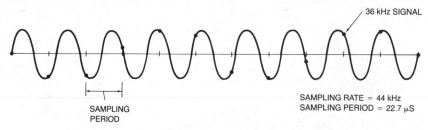

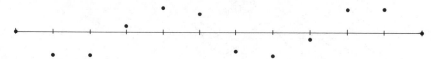

(A) A 36 kHz waveform is sampled at 44 kHz.

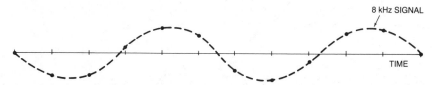

(B) Samples are stored.

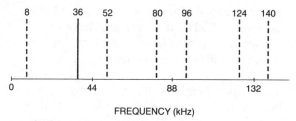

(C) Upon reconstruction the 36 kHz signal will be filtered out leaving an aliased 8 kHz signal.

Fig. 2-6. **Aliasing is a condition disallowed by the sampling theorem.**

is gone, but the 8 kHz signal remains, containing samples as innocuous as any legitimate 8 kHz signal. What does that unwanted signal do to the fidelity of our audio system? Distortion.

2.2-2 Image Aliasing • There are other manifestations of aliasing. Alias components occur not only around the sampling frequency, but also in the multiple images produced by sampling, as illustrated in the spectrum in Fig. 2-7. For

Fig. 2-7. **Alias image spectrum**

example, all of these components would be produced in an aliasing scenario: $\pm S \pm F$, $\pm 2S \pm F$, $\pm 3S \pm F$, etc. Thus, in our example of a 44 kHz sampler and a 36 kHz input signal, some of the resulting frequencies would be: 8, 52, 80, 96, 124, 140 kHz, etc. Although only the $S - F$ component will bother us directly as an interfering frequency in the audio bandwidth, the sampling images will continue to affect the audio bandwidth, no matter how high frequency F becomes.

Aliasing is about as bad as being trapped in a house of mirrors. If you were sampling at 44 kHz, an input frequency from 0 to 22 kHz would sound fine, but as the frequency ranged from 22 kHz to 44 kHz, we would hear it returning

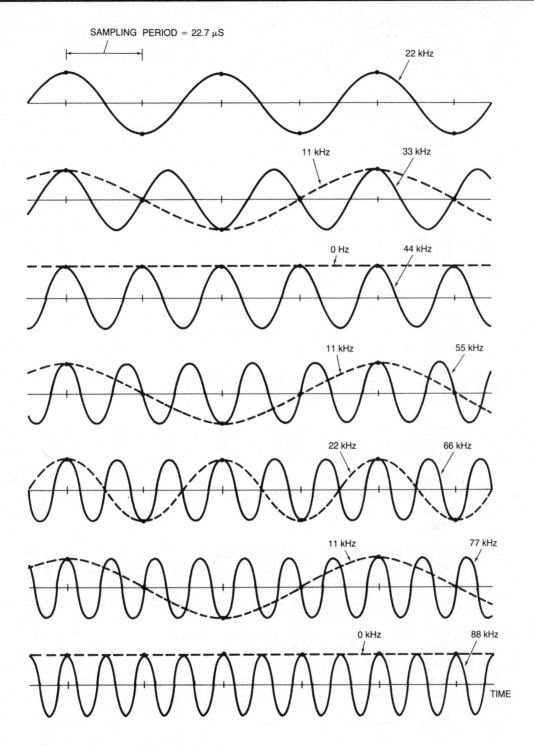

Fig. 2-8. **Aliased frequencies in the audio band.**

as a frequency descending from 22 kHz to 0. If the input frequency was raised from 44 kHz to 66 kHz, it would appear again from 0 to 22 kHz, etc. Fig. 2-8 shows how different aliased frequencies appear in the audio band from 0 to 22 kHz as the input frequency increases.

2.2-3 Harmonic Aliasing • More complex tones exaggerate the problem. Our simple sine tone examples have limited foldover to the one and only partial of a sine wave. With complex tones, aliasing frequencies could be generated separately for each harmonic. For example, the second harmonic of a complex waveform with an 11 kHz fundamental would be 22 kHz which would be critically sampled by a 44 kHz sampler. However, the third harmonic at 33 kHz would be aliased back at 11 kHz to add to the fundamental. The sixth harmonic at 66 kHz would alias at 22 kHz to add to the second harmonic. Obviously, we would have our hands full with misrepresented partials. More typically, the fundamental and its harmonics would not be submultiples of the sampling frequency. For example, as shown in the spectrum of Fig. 2-9, a complex fre-

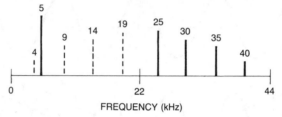

Fig. 2-9. **Harmonic aliasing.**

quency of 5 kHz would produce a fifth harmonic at 25 kHz, which would alias at 19 kHz, the sixth harmonic at 30 kHz would alias at 14 kHz, the seventh harmonic at 35 kHz would alias at 9 kHz, and the eighth harmonic at 40 kHz would alias at 4 kHz, just below the fundamental. Again, distortion would result. Under most conditions the effect would be almost impossible to hear because of the low amplitudes of most harmonics, and masking by the music itself.

2.2-4 Solution to Aliasing • As bad as aliasing might be, in practice, it isn't a serious problem. In fact, in a well designed digital recording system, aliasing will be prevented from occurring. The solution is simple, we merely band limit the input frequencies, with a sharp low-pass filter, sometimes called the anti-aliasing filter, designed to provide large attenuation at half the sampling rate to make sure the throughput never exceeds half that frequency, as shown in Fig. 2-10. An ideal filter, as shown in Fig. 2-11, would have a "brickwall" characteristic and infinite attenuation; however, in practice the filter cannot be so steep and is designed to provide attenuation only beneath the amplitude resolution of the system thus disallowed frequencies will never become digitized. Neither

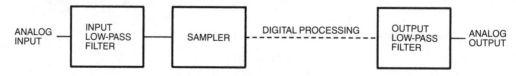

Fig. 2-10. **To prevent aliasing an input low-pass filter precedes all digitization systems.**

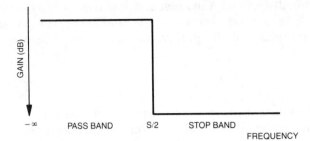

Fig. 2-11. **Ideal low-pass filter characteristic.**

disallowed fundamentals nor partials are allowed to enter the sampler, thus aliasing cannot occur. If those frequencies don't exist, neither will aliasing. It is thus critical to observe sampling theory, and low-pass filter the input signal to a digitization system. If aliasing were allowed to occur, there is no technique able to remove the aliased frequencies from the original audio bandwidth. As we shall see, extremely low level harmonic aliasing can occur after the anti-aliasing filter because of quantization error. A noise signal called dither is used

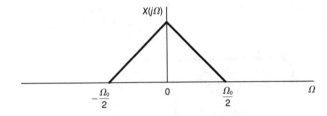

(A) Fourier transform of original analog signal.

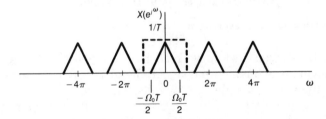

(B) Fourier transform of band-limited discrete-time signal obtained by sampling.

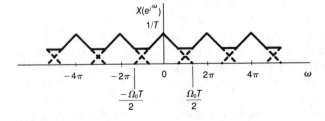

(C) Spectra overlap results in aliasing.

Fig. 2-12. **Analysis of frequency spectra shows effects of aliasing.**

to eliminate this distortion. A mathematical analysis of frequency spectra, as shown in Fig. 2-12, summarizes the effects of aliasing. The dotted line in Fig. 2-12B shows the effect of the output low-pass filter to recover the original spectra.

2.3 Quantization

To record an audio signal, two dimensions of information must be stored. Sampling implicitly saves time information and quantization saves amplitude information. Quantization is thus the measured value of the analog signal at sample time. With quantization, as with the measurement of any analog event, accuracy is limited by the system's resolution. Because of finite word length, a digital system's resolution is limited, thus a measuring error is introduced. This error is similar to noise in an analog system; however, it differs because its character changes with signal amplitude.

2.3-1 Analog and Digital Approximation • Let's use an example to illustrate the effects of quantization and differentiate its error from the error inherent in an analog system. Suppose that we have connected two voltmeters as shown in Fig. 2-13, one analog and one digital, to a recording console and at the final

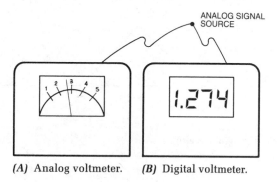

(A) Analog voltmeter. (B) Digital voltmeter.

Fig. 2-13. **Approximation in measurement.**

chord of the Beethoven Ninth, we read both meters, measuring the voltage corresponding to the acoustic input signal. Given a good meter face and a sharp eye, we read the analog needle at 1.27 volts. The digital meter, a rather cheap model, has only two digits and thus we read 1.3 volts; if we had paid a little more for a three digit meter we might have read 1.27 volts and a four digit meter might have read 1.274 volts. Now, both types of meters are always in error. The error in the analog meter is because of the ballistics of the mechanism and our difficulty in reading the meter. Even under ideal conditions, at some point, any analog measurement capacity is lost in the device's own noise.

With the digital meter, the nature of the error is different. Accuracy is limited by the resolution of the meter, that is, by the number of digits displayed. The more digits, the greater the accuracy, but the last digit will always round off relative to the actual value, for example 1.27 was originally rounded off to 1.3. Under the best conditions the last digit would be completely accurate; for example, a voltage of exactly 1.3000 would be shown as 1.3, and under the worse condition the rounding off will be one-half increment away; for example, 1.250 would be rounded off to 1.2 or 1.3. If a binary system is being used for the measurements, we say that the error resolution of the system is one-half the least significant bit (LSB). For both analog and digital systems, the problem of measuring an analog phenomenon such as amplitude leads to error. The philosophical question of which is better, analog or digital, is usually decided in the marketplace. At least as far as voltmeters is concerned, for equal or less cost, we can build digital meters with greater resolution, ease of use, and reliability. A digital read-out is an inherently more robust kind of measurement; we gain more information about an analog event when it is characterized in terms of digital data. The analog voltmeter has gone the way of the slide rule.

Quantization is thus the technique of incrementing an analog event to form a discrete number. Of course, a digital system usually dictates the use of a binary number system; in terms of the quantizing hardware, the number of increments is determined by the length of the data word, that is, the number of bits available to form the representation. Just as the number of digits in our digital voltmeter determined our resolution, the number of bits in our digitization equipment determines resolution. As we shall see, that decision is primarily influenced by the cost of the analog-to-digital (A/D) converter; longer word length A/Ds are quite expensive.

2.3-2 Approximation in Measurement • The task of recording and reproducing music can be simply summarized—we want to form a representation of the music. The closer our representation is to the original, the better. Unfortunately, reality is stubborn in its ability to defy re-creation, and we are left with a challenging endeavor as we attempt to create an approximation of the original event by saving as much information as possible. The essential problem lies in the complexity of even the simplest acoustical waveform and the dual nature of the information it carries. No matter which recording system we employ, to characterize an acoustical event, we must show correlated time and amplitude information. Thus, a vinyl LP has a groove, the length of which implicitly encodes time, and lateral variations which encode amplitude. In a digital system, both time (implicitly again) and amplitude are stored as discrete pieces of information. We have discussed sampling, a method of periodically taking a measurement, of course, taking a measurement of a varying event is meaningful only if both the time and the value of the measurement are stored. Sampling represents the time of the measurement, and quantization represents the value of the measurement, or in the case of audio, the amplitude of the waveform at sample time. Sampling and quantization are thus the fundamental components of digitization, and together, at least in theory, can characterize any acoustical event. Both sampling and quantization become variables which determine respectively the bandwidth and resolution of the approximation. An originally analog waveform may be mapped into a series of pulses, the amplitude of each will yield a number which represents the analog value at that instant. The interplay between sampling rate and quantization is shown in Fig. 2-14. Correct sampling of a band limited signal is a lossless process, but choosing the amplitude value at sample time is not. Any choice of scales or codes, as shown in Fig. 2-15, results in the same realization that incremental mapping can never totally encode a continuous analog function. An analog waveform has an infinite number of amplitude values whereas we can only choose from a finite number of increments so our chosen value will be only an approximation to the actual. In other words, with quantization, there is an error.

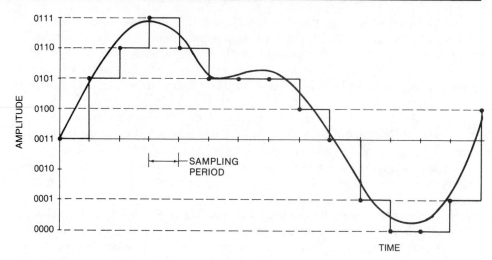

(A) Number of quantization intervals is low, resulting in a poor approximation.

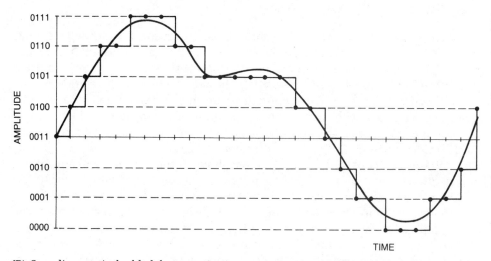

(B) Sampling rate is doubled, but quantization approximation is still poor.

Fig. 2-14. **Four examples comparing different**

2.3-3 Signal-to-Error Ratio • With a binary number system, the word length determines the number of quantizing increments available; this can be computed by raising the word length to the power of 2, as shown here:

Number (N) of quantization intervals in a binary word: $N = 2^n$ where n is the number of bits in the word.

$2^1 = 2$	$2^6 = 64$	$2^{11} = 2048$	$2^{16} = 65536$
$2^2 = 4$	$2^7 = 128$	$2^{12} = 4096$	$2^{17} = 131072$
$2^3 = 8$	$2^8 = 256$	$2^{13} = 8192$	$2^{18} = 262144$
$2^4 = 16$	$2^9 = 512$	$2^{14} = 16384$	$2^{19} = 524288$
$2^5 = 32$	$2^{10} = 1024$	$2^{15} = 32768$	$2^{20} = 1048576$

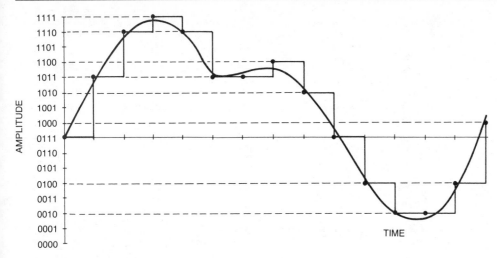

(C) With low sampling rate, but double the number of quantization intervals, quantization approximation is improved.

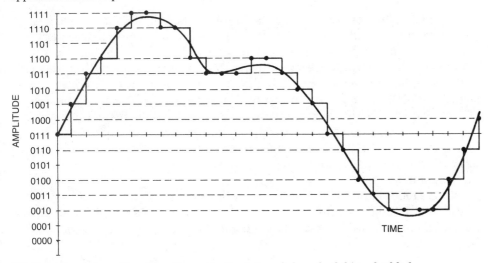

(D) Sampling rate and number of quantization intervals have both been doubled.

quantization and sampling resolution.

Thus, an 8 bit word, or byte, would accommodate $2^8 = 256$ increments, a 16 bit byte would map $2^{16} = 65,536$ increments. The more bits the better the approximation, but as we have seen, there must always be an error associated with quantization because the limited number of amplitude choices contained in the binary word can never completely map an infinite number of analog possibilities. No matter how many increments are available, there can always be an analog amplitude in between. At some point, the quantizing error becomes audibly indistinguishable; most manufacturers have agreed that 16 bits provide an adequate representation. However, that doesn't rule out longer data words in the future, or the use of gain control circuits to optimize quantization and thus reduce error level.

ANALOG VOLTAGE	2's COMPLEMENT BINARY	OFFSET BINARY	SIGN & MAGNITUDE BINARY
+1.00			
	0111	1111	0111
.875			
	0110	1110	0110
.750			
	0101	1101	0101
.625			
	0100	1100	0100
.500			
	0011	1011	0011
.375			
	0010	1010	0010
.250			
	0001	1001	0001
.125			
	0000	1000	0000
.000			
	1111	0111	1000
−.125			
	1110	0110	1001
−.250			
	1101	0101	1010
−.375			
	1100	0100	1011
−.500			
	1011	0011	1100
−.625			
	1010	0010	1101
−.750			
	1001	0001	1110
−.875			
	1000	0000	1111
−1.00			

Fig. 2-15. **Quantization mapping.**

Word length determines the resolution of our digitization system, and hence provides an important specification to measure the system's performance. Sometimes our chosen increment will be exactly at the analog value, usually it will not be quite exact. At worse, the analog level we desire to encode will be one-half increment away, that is, there will be an error of half the least significant bit of the quantization word. For example, suppose the binary word 101000 maps the analog increment of 1.4 volts, and 101001 maps 1.5 volts, and the actual analog value at sample time is unfortunately 1.45 volts, as shown in Fig. 2-16. Since 101000½ is not available, we would have to round up to 101001 or down to 101000, either way, we would be in error by one-half of an increment magnitude.

In characterizing digital hardware performance we may formulate a ratio of the total number of intervals (N-1) covered by our quantization scheme to the maximum interval error, as shown here:

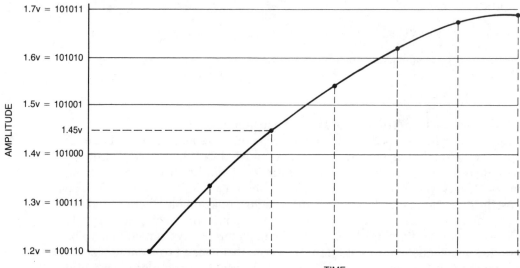

Fig. 2-16. **Quantization error is limited to one-half LSB.**

the approximate signal-to-error ratio $= \dfrac{\text{maximum number}}{\text{maximum error}}$

for example, with 16 bits:

approximate signal-to-error ratio $= \dfrac{65,535}{.5}$

$= 131,070$ which is equivalent to 98 dB

This ratio of maximum expressible amplitude to error determines the signal to error ratio of the system; the signal to error (S/E) ratio specification of a digital system is closely akin, but not identical in nature, to the signal to noise (S/N) ratio of an analog system. The S/E relationship can be more rigorously derived using a ratio of signal to error voltage levels, as shown in Fig. 2-17.

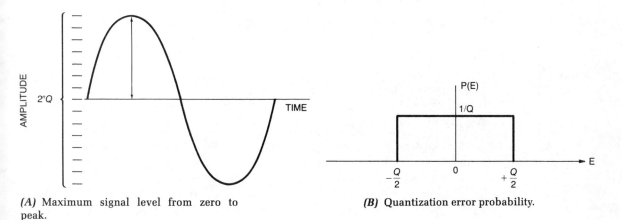

(A) Maximum signal level from zero to peak.

(B) Quantization error probability.

Fig. 2-17. Signal-to-error ratio.

More specifically, signal to error ratio can be expressed by calculating first a maximum signal value, then an error value, then forming their ratio.

Given a quantization system in which each interval is "Q" amplitude, and "n" is the number of bits, the maximum signal level from zero to peak is:

$$\frac{1}{2}\,(2^{n}\,Q) = 2^{n-1}\,Q$$

and the maximum rms signal value is thus:

Signal Voltage (rms) $= \dfrac{2^{n-1}\,Q}{\sqrt{2}}$

The error signal may be analyzed by examining the quantization error probability function.

When the input signal has high amplitude and wide spectrum, the error has an equal probability of being any value between $+\,Q/2$ and $-\,Q/2$ where Q is one quantization interval. This is illustrated in Fig. 2-17B by a random probability density function. Since the error is random from sample to sample, the error spectrum is flat.

The energy contained in the noise can be calculated by integrating all error values obtained by multiplying the energy in one error (E) by its probability P (E) dE:

$$\text{error energy} = \int_{-Q/2}^{+Q/2} E^2 \, P(E) \, dE \qquad \text{with } P(E) = \frac{1}{Q}$$

which yields:

$$\text{Error voltage (rms)} = \frac{Q}{\sqrt{12}}$$

in terms of signal-to-error ratio: $S/E = \dfrac{\dfrac{2^{n-1} Q}{\sqrt{2}}}{\dfrac{Q}{\sqrt{12}}} = 2^n \sqrt{1.5}$

thus signal-to-error ratio $= 2^n \sqrt{1.5}$ where n is number of bits which in decibels is:

signal-to-error ratio = 6.02 n + 1.76 dB.

For example, 16 bits again yields about 98 dB S/E whereas, a 14 bit system is slightly inferior at 86 dB.

2.3-4 Quantization Error • Quantization error is the difference between the actual analog value at sample time, and the chosen quantization interval value, as shown in Fig. 2-18. At sample time, the amplitude value must be chosen from the nearest quantization interval. At best (sample points 11 and 12 in the figure) the waveform coincides with quantization intervals. At worst (sample point 1 in the figure) the waveform is exactly between two intervals. Quantization error, the difference between the actual and measured values, is thus limited to $\pm\frac{1}{2}$ interval at sample time. At the system output, when the sample values are used to recreate an analog waveform, this error will be contained in the output signal. Perceptually, quantization error is similar to analog white noise; however, the noise floor of an analog system differs from the error floor of a digital system. For example, in a digital system, when there is no signal present there is no error or noise whereas noise is always uniformly present in an analog system. The perceptual qualities of quantization error differ from analog noise because the quantization error varies with the amplitude of the input signal; specifically, quantization error is perceptively changed from noise to distortion at low amplitude input signals.

2.3-5 Other Quantization Methods • Quantization is more than just word length, it is also a question of hardware design and formats. There are many techniques available to accomplish quantization, and different strategies determine how the analog signal is mapped onto the increments. For example, we could use a linear or nonlinear distribution of quantization intervals along the amplitude scale. Alternatively, a delta modulation system could be used in which only a one bit quantizer is used to encode amplitude, using the single bit as a sign bit. Those algorithm decisions influence the efficiency of the available bits as well as the relative effects of the error. For example, a linear

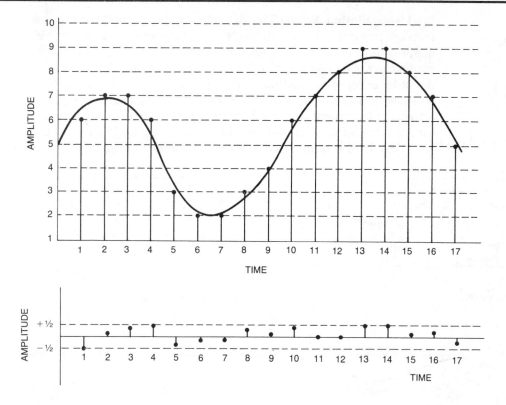

Fig. 2-18. **Quantization error at sample time.**

quantizer produces a relatively high error with low-level signals which span only a few increments. A nonlinear system such as a floating point converter could be used to amplify low-level signals to utilize the fullest possible incremental span. While that improves the overall signal to error ratio, the noise modulation by-product is undesirable and requires special masking. Manufacturers have examined the trade-offs of different quantization systems, and generally determined that a fixed, linear quantization scheme is most suitable for music recording.

2.4 Dither

It is the nature of quantization to introduce quantization error. Although the error occurs at a very low level, its presence must be considered in a high fidelity music system. With large amplitude complex signals there is little correlation between the signal and the error, thus the error is random, and sounds similar to analog white noise. With low-level signals, the character of the error changes as it becomes correlated to the signal; small yet measurable distortion results. To remove this correlation, an analog noise signal is often added to the audio signal prior to the sampler; this randomizes the effects of the quantization error. This signal, called dither, does more than mask the quantization noise; it can provide the digital system with the ability to encode amplitudes smaller than the least significant bit, in much the same way that an analog system can retain signals below its noise floor.

2.4-1 Granulation Noise • The discrete nature of digital audio permits us to employ powerful digital methods for processing and storage. However, in quantization, the discrete nature of digital and the ensuing amplitude increments can create nonlinearity. Consider the case of a signal with amplitude on the order of one quantization increment, as shown in Fig. 2-19. It would either move within the increment, resulting in a dc signal, or cross back and forth across the increment threshold, resulting in a square wave output signal. That square wave, created at very low levels, suggests that quantization ultimately acts like a hard limiter, in other words, severe distortion takes place.

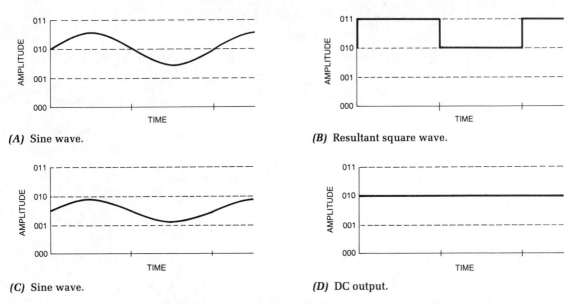

(A) Sine wave.

(B) Resultant square wave.

(C) Sine wave.

(D) DC output.

Fig. 2-19. **Quantization error on low-level signal.**

To make matters worse, the square wave created by quantization is rich in odd harmonics, extending far beyond the Nyquist (half sampling) frequency, thus aliasing can appear *after* the anti-aliasing low-pass filter. The square wave appears after the sampler, but it is effectively sampled; as Blesser notes, the effect of sampling the output of a limiter or limiting the output of a sampler are indistinguishable. This aliasing accentuates the quantization error by creating a spectrum of error components. That quantized fluctuation in the noise floor is perceived as a particularly nasty kind of sound called granulation noise because it has been described as a gritty sound.

Consider the example of a low-level square wave of 15.333 kHz in a 44 kHz sampling system, as shown in the spectrum in Fig. 2-20. The third harmonic creates a 1.999 kHz component, the fifth harmonic an 11.333 kHz component, and the seventh harmonic a 19.333 kHz component, etc. An entire quantization error spectrum has been created; these are entirely new components not masked by the original fundamental or its harmonics. If the harmonics are very close to a multiple of the sampling frequency, the beat tones drift through the frequency origin to produce a sound called "birdsinging" or "birdies." The anti-aliasing filter is powerless against these harmonics since they are created in the signal chain after the filter.

For broadband high amplitude input signals (such as typically found in

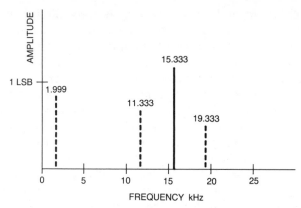

Fig. 2-20. **Quantization error spectrum.**

music) the quantization error is perceived similarly to white noise. However, the perceptual qualities of the error are less benign for low-amplitude signals and high level signals of very narrow bandwidth. This is similar to the fact that white noise is perceptually benign because successive values of the signal are random, whereas predictable noise signals are more perceptible. For broadband high-level signals the statistical correlation between successive samples is very low; however, it increases for broadband low-level signals and narrow bandwidth high-level signals. As the statistical correlation between samples increases, error initially perceived as benign white noise becomes a more complex kind of distortion.

2.4-2 Effects of Dither • A digitization system must suppress any audible qualities of its quantization error. Obviously, the number of bits in the quantizing word could be increased, with a resultant decrease in error amplitude of 6 dB per additional bit. This is uneconomical, and it is not known how many extra bits would be needed to reduce the audibility of quantization satisfactorily. An alternative is to add a small amount of analog white noise to the input signal;

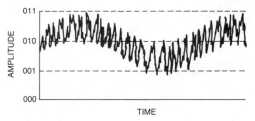

(A) Dither is added to the sine wave input.

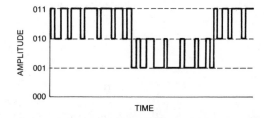

(B) Sine wave is preserved in pulse width modulation.

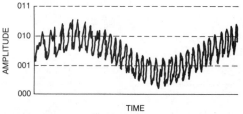

(C) Dither is added to the sine wave input.

(D) Sine wave is preserved in pulse width modulation.

Fig. 2-21. **Quantization error and the effects of dither.**

this noise is referred to as dither. Dither has been employed in designs since its use in video technology circa 1950, and is now being applied to digital audio technology. With dither, a small amount of noise removes the quantization artifacts from a signal; importantly—it doesn't mask the artifacts, it removes them. Let's reconsider our example of a signal crossing a quantization threshold. But this time let's add a dither signal equal to a quantizing step. The result shown in Fig. 2-21 is a pulse signal which preserves the information of the original signal. The quantized signal switches up and down as the dithered input varies, tracking the input's average value. This information is encoded in the varying width of the digital signal's pulses. This kind of information storage is known as pulse width modulation and it accurately preserves the input signal's waveform. The average value of the quantized signal can move continuously between two levels thus the incremental dangers of quantization have been alleviated. Audibly, the result is the original waveform, with added noise. But that is far more desirable than the severely distorting square wave.

Vanderkooy and Lipshitz have demonstrated this technique with the example of a 1 kHz sine wave with a peak to peak excursion of one-half LSB, shown in Fig. 2-22. The familiar square wave is output from the digital-to-analog converter as the limiting action takes place. But when dither with an

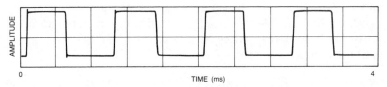

(A) A 1 kHz sine wave with amplitude of one-half LSB is quantized to produce a square wave.

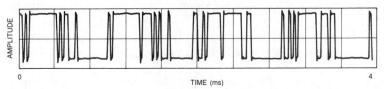

(B) Dither of one-third LSB rms amplitude has been added to the sine wave before quantization.

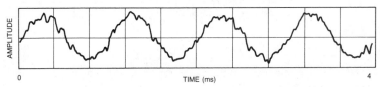

(C) Modulation carries the encoded sine wave information, as can be seen after 32 averagings.

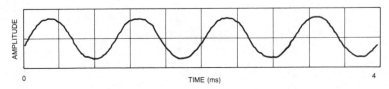

(D) Modulation carries the encoded sine wave information, as can be seen after 960 averagings in (D).

Fig. 2-22. Effects of dither.

amplitude of one-third LSB is added, a pulse width modulated waveform results. While this might still look distinctly digital, its hidden sine wave information can be revealed when the signal is averaged. A fairly clean sine wave distinctly emerges. And that averaging technique isn't some kind of trick; rather, it illustrates how the ear responds in its perception of acoustical signals.

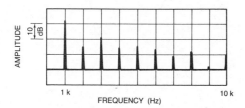

(A) Without dither.

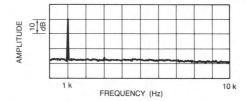

(B) With dither.

Fig. 2-23. **Effect of dither on harmonic distortion.**

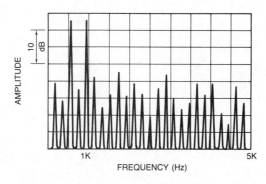

(A) Without dither.

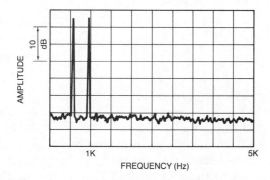

(B) With dither.

Fig. 2-24. **Effect of dither on intermodulation distortion.**

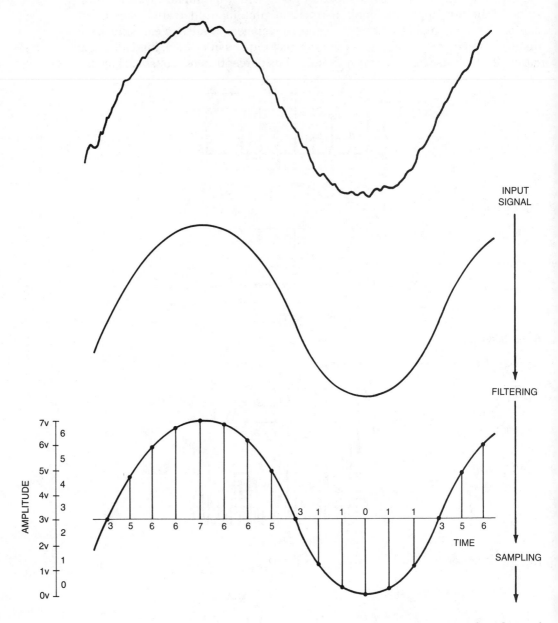

INPUT
SIGNAL

FILTERING

SAMPLING

TIME

AMPLITUDE

7v
6v
5v
4v
3v
2v
1v
0v

6
5
4
3
2
1
0

3 5 6 6 7 6 6 5

3 1 1 0 1 1

3 5 6

Fig. 2-25. **Essential workings of**

Our ears are quite good at resolving narrow band signals below the noise floor because of the averaging properties of the basilar membrane. According to Vanderkooy and Lipshitz, the ear behaves like a one-third octave filter with a narrow bandwidth such that the quantization error, which has been given a white noise character by dither, is averaged by the ear and the original narrow band sine wave is heard without distortion. In other words, dither changes the digital nature of the quantization error into a white noise, and the ear may then resolve signals with levels well below one quantization level.

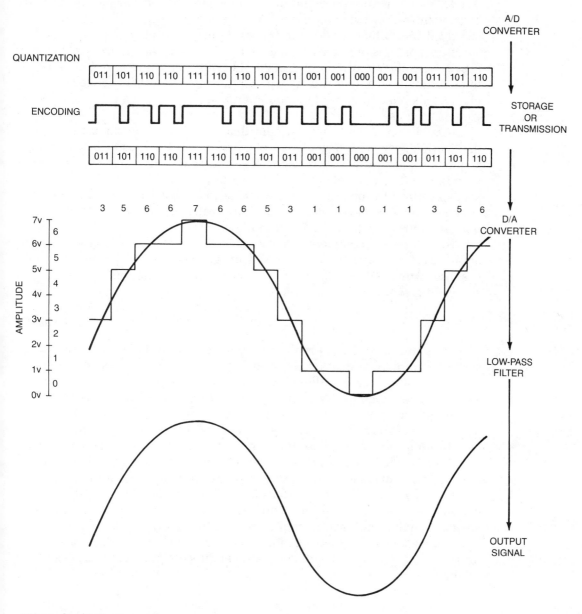

a PCM audio digitization system.

This is an important conclusion. With dither, the resolution of a digitization system is below the least significant bit. By encoding the audio signal with dither to produce modulation of the quantized signal, we may recover that information, even though it might be smaller than the smallest increment of the quantizer. Furthermore, dither minimizes distortion caused by quantization by reducing those artifacts to white noise. Proof of this is shown by harmonic and intermodulation distortion measurements with and without dither, Vanderkooy and Lipshitz have demonstrated this benefit of dither, using dither on the order of one-third quantization increment, as shown in Figs. 2-23 and 2-24. A 1 kHz sine wave with amplitude of one-half LSB becomes a square wave after quantization with the harmonic spectrum shown in Fig. 2-23A. When one-third LSB dither is added to the signal before quantization, the resultant spectrum contains only the original signal, and wide-band noise. As shown in Fig. 2-24A, signals of 600 and 1000 Hz with amplitude of 1 LSB produce intermodulation components. When one-third LSB of dither is added, the original signals remain, and intermodulation components have been replaced by wide-band noise.

In addition to its beneficial contributions, dither contributes noise to a digitization system. A degradation of a few dB in the broadband noise floor results, an increase which is negligible compared to the large signal to error ratio inherent in a digital system. Dither thus represents a unique and unlikely concept. The idea of adding analog noise to a digital system seems a little strange. But as we have seen, the result is greater resolution and lower distortion, and thus higher fidelity for the audio system.

For the sake of completeness, we might note that the dictionary definition of dither is "a highly nervous, confused, or agitated state." In minute quantities, it successfully makes a digitization system a little more analog in the good sense of the word.

Summary

Sampling and quantizing are the two fundamental criteria for a digitization system. Sample rate determines band limiting and thus frequency response, and quantization is a question of word length, which determines the dynamic range of the system, measured by signal to error ratio. Although band limited sampling is a lossless process, quantization is one of approximation. Aliasing occurs when band limiting is not observed, and dither minimizes the effects of quantization. In general, a sampling rate of 44.1 kHz or 48 kHz, and a word length of 16 bits yields fidelity comparable to or better than the best analog systems, with advantages such as longevity and fidelity of duplication.

In preparation for a thorough examination of a complete digitization system in Chapters 3 and 4, we may summarize the process of transforming from analog to digital information, and then back to analog, as shown in Fig. 2-25. The audio signal is sampled, quantized, converted to binary form, and encoded for recording or transmission. Reversing the process produces a replica of the original signal.

Chapter 3

Digital Audio Recording

Introduction

The hardware design of a digital audio recorder embodies the fundamental conceptual principles such as sampling and quantizing. The digital audio system is designed to accomplish those processing tasks, thus the analog signal is sampled, converted to digital numerical form, and conditioned in preparation for its digital storage. Subsystems such as anti-aliasing filter, sample and hold, analog-to-digital converter, modulator and dither circuits constitute the hardware encoding chain of a digital audio recorder. Similarly, audio processing devices such as digital delay lines and digital reverberators encode the analog waveform in preparation for signal processing. Although other architectures have been designed, the linear pulse code modulation (PCM) system is the most illustrative of the nature of audio digitization and indeed the most widely used system. This chapter, and that following, describe the hardware architecture best suited to accomplish a PCM digitization system. Such a system could accomplish the essential pre- and post-processing for either a digital recorder, or real-time digital processor.

3.1 Pulse Code Modulation

In theory, there are an almost endless number of techniques available to perform the task of digitally encoding audio signals. They are all fundamentally identical in their operation of transforming analog signals into digital data but in practice they differ widely in their relative efficiency in terms of required bandwidth, signal-to-noise ratio, and accuracy. There is only one technique which has found widespread acceptance in audio digitization, and although other techniques presently exist, and newer ones will be devised, they will measure their success against that of the pulse code modulation digitization system.

3.1-1 Modulation Systems • Modulation is nothing more than a means of encoding information for the purpose of transmission or storage. Techniques such as amplitude modulation (AM) and frequency modulation (FM) have long been used to modulate carrier frequencies with analog audio information for radio broadcast. Since these are continuous kinds of modulation, they are referred to as wave parameter modulation. They are illustrated in Fig. 3-1. With digital information, the possibilities of other forms of modulations present them-

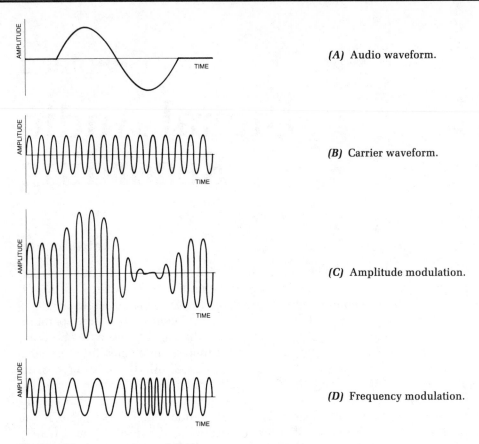

(A) Audio waveform.

(B) Carrier waveform.

(C) Amplitude modulation.

(D) Frequency modulation.

Fig. 3-1. **Wave parameter modulation.**

selves; pulse parameter modulation schemes such as pulse amplitude modulation (PAM), pulse number modulation (PNM), pulse width modulation (PWM), pulse position modulation (PPM), and pulse code modulation (PCM) all offer ways to encode, transmit and store digital data. With PAM, PWM, and PPM, shown in Fig. 3-2, variations in pulse amplitude, or time position, width is used to represent the analog signal's value at sample time. These employ hybrid pulse techniques in which the amplitude, position, or width of the pulse encodes the information directly. With PNM and PCM, shown in Fig. 3-3, a pulse or code is directly produced from the input signal; for audio encoding these are more efficient techniques. PNM generates a string of pulses, the pulse count thus specifies the amplitude of the input signal. However, for high resolution, a large number of pulses is required. With PCM the pulse chain has been coded to greatly reduce the number of pulses, as well as the bandwidth required to store the data. Overall PCM has been judged to be the most effective form of digital representation for high fidelity audio signals.

3.1-2 PCM • Architecture for a PCM system closely follows an easily conceptualized means of designing a digitization system; the input waveform is sampled, and the value of each sample level is converted into binary coded notation. A bit stream of amplitude data is formed in which one or more channels is multiplexed, that is, merged into one data stream to form a single serial channel, ready for recording. The PCM process is depicted in Fig. 3-4,

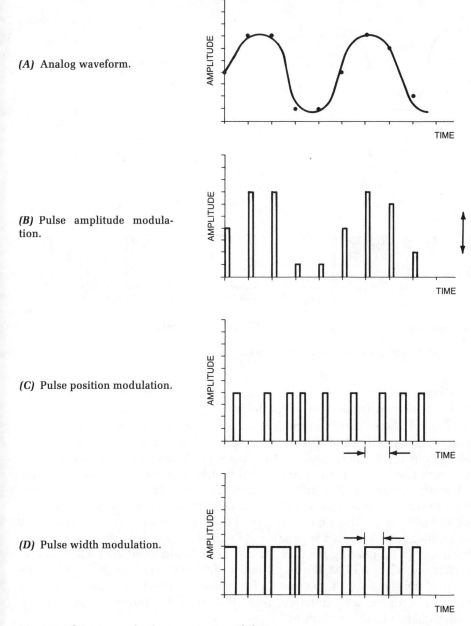

(A) Analog waveform.

(B) Pulse amplitude modulation.

(C) Pulse position modulation.

(D) Pulse width modulation.

Fig. 3-2. **Three types of pulse parameter modulation.**

showing the transformation of the audio signal from analog value to PCM data. The original analog waveform is time sampled and its amplitude quantized by the analog-to-digital (A/D) converter. Binary numbers are sent to the storage medium as a series of pulses representing amplitude. If two channels are to be sampled, the PCM data may be multiplexed to form one data stream. Since a binary code is being created directly, this data is easily recognized by the system computer for computation such as signal processing and error correction. Following storage, upon playback, the bit stream is decoded to recover

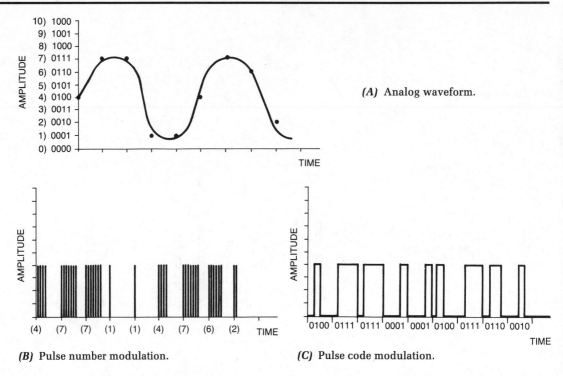

(A) Analog waveform.

(B) Pulse number modulation.

(C) Pulse code modulation.

Fig. 3-3. **Numerical pulse parameter modulation.**

the original amplitude information, at proper sample times, and the analog waveform is reconstructed by a digital-to-analog (D/A) converter.

In any PCM system, the binary word length determines the number of quantization increments available to encode the analog amplitude. In a linear PCM system the 2^n increments of the analog increment scale are all of equal height and the intervals are given digital words in monotonic order. In other PCM systems, such as floating point, a nonlinear method divides the digital word into an exponent and mantissa for expanded dynamic range. A greatly modified form of PCM called delta modulation, in which only one bit is used for quantizing, is discussed separately at the end of Chapter 4. For comparison, fixed linear PCM, floating point PCM, and delta modulation are shown in Fig. 3-5. With fixed linear PCM, the amplitude scale remains unchanged. With floating point, the scale varies, decreasing for greater accuracy as signal amplitude decreases. With delta modulation, one bit is used to record the positive or negative change from the prior bit.

The recording section for a classic stereo PCM digitization scheme consists of input amplifiers, a dither generator, input low-pass filters, sample and hold circuits, analog-to-digital converters, a multiplexer, digital processing and modulation circuits, and a storage medium such as digital tape. An encoding section block diagram is shown in Fig. 3-6. This hardware collection is thus a realization of our previous conceptual mathematical theorems. As we have discussed, an audio digitization system is really nothing more than a kind of transducer which processes the audio signal for digital storage, then processes it again for reproduction. While that sounds simple, the hardware must be

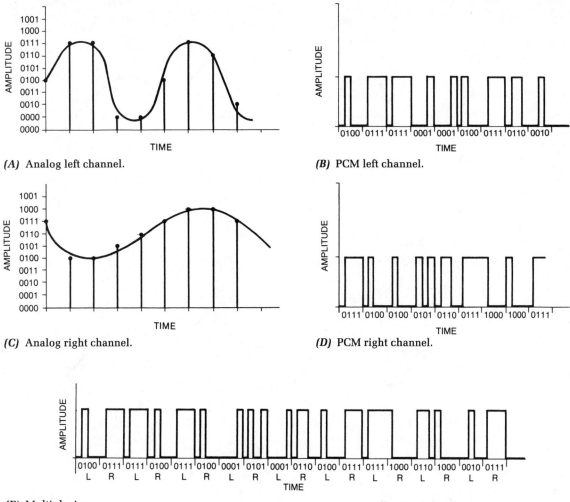

(A) Analog left channel.

(B) PCM left channel.

(C) Analog right channel.

(D) PCM right channel.

(E) Multiplexing.

Fig. 3-4. **PCM encoding and multiplexing.**

carefully engineered. Its success in accomplishing its task, and the quality of the reproduced audio, depend entirely on the quality of the system's design. Each subsystem must be carefully considered.

Aside from the requirements of absolute quality such as low distortion and high signal-to-noise ratio, so as to not compromise the fidelity of the ensuing digital system, the input analog amplifiers have no special design features from a digital standpoint. The first engineering challenge occurs with the dither generator.

3.2 Dither Generator

Dither, as we have seen, is the analog noise signal added to the analog signal; it is used to remove the artifacts of quantization error. The dither signal causes the audio signal to constantly move between quantization levels; this averages the quantization differences and indeed encodes information with resolution

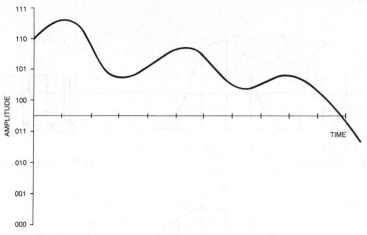

(A) Fixed linear PCM.

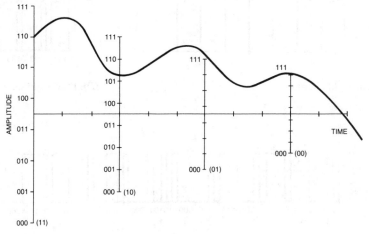

(B) Floating point.

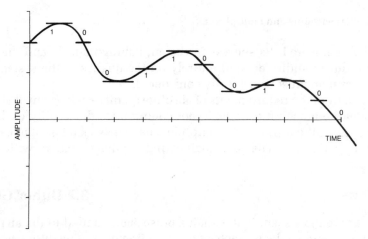

(C) Delta modulation.

Fig. 3-5. **Three conversion methods.**

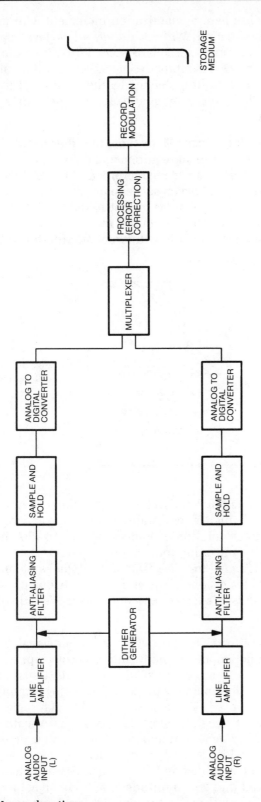

Fig. 3-6. **Linear PCM record section.**

less than the amplitude of a quantization increment with modulation similar to pulse width modulation. Without dither, a low level signal would be encoded by an A/D converter as a square wave. With dither the output of the A/D is the signal with noise. Perceptually, the effects of dither are much preferred because noise is more readily tolerated by the ear than distortion. The generation of the distortion caused by quantization error is thus prevented by the addition of dither.

3.2-1 Requirements for Dither • Many types of dither signals exist, but for audio applications two mathematical requirements define the most useful dither signal; the signal should have a rectangular probability density function, and there should be no correlation between successive dither samples. This ensures that the resulting noise will be similar to that found in analog systems; which is least disturbing perceptually. The amplitude of the dither signal is important. Fig. 3-7 shows the effects of adding different amplitudes of dither to the audio

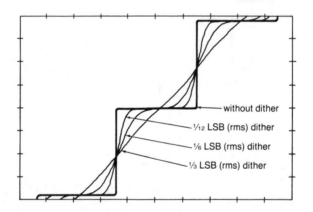

without dither

⅟₁₂ LSB (rms) dither

⅙ LSB (rms) dither

⅓ LSB (rms) dither

Fig. 3-7. **Effects of dither with different amplitudes.**

signal, as demonstrated by Vanderkooy and Lipshitz; quantization artifacts are removed as relatively higher amplitudes of dither are added. It can be seen that a dither signal with an amplitude of one-third quantization increment successfully smooths the quantization increments. Of course, too much dither would overly decrease the signal-to-noise ratio of the digital system. In general, dither with amplitude between one-sixth and one-third quantization increment is employed; this results in a decrease in the system S/N ratio of less than 1.5 dB.

3.2-2 Dither Generation • The most effective dither signal, a rectangular probability density function of constant and precise amplitude, is difficult to achieve thus designers employ other dither signals. Gaussian (white) noise is often used; it is easy to generate with common analog techniques, for example, a zener diode can be used as a noise source. Although some noise modulation is added to the audio signal, its effect is very slight. A mathematical comparison of the probability density functions of rectangular density and Gaussian density is shown in Fig. 3-8. Other designers believe that the analog noise inherent in the input amplifier and low-pass filter is an adequate dither signal. It has been observed that the amplitude of a dither signal may be decreased if a sine or square wave just below the Nyquist frequency with an amplitude of

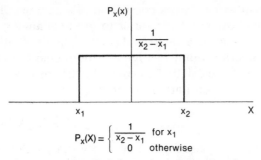

$$P_x(X) = \begin{cases} \dfrac{1}{x_2 - x_1} & \text{for } x_1 \\ 0 & \text{otherwise} \end{cases}$$

(A) Rectangular density function is the ideal choice for a dither signal.

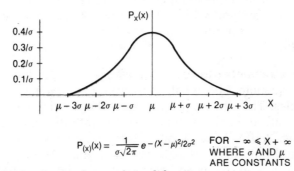

$$P_{(x)}(x) = \frac{1}{\sigma\sqrt{2\pi}} e^{-(X-\mu)^2/2\sigma^2} \quad \begin{array}{l} \text{FOR } -\infty \leqslant X+ \infty \\ \text{WHERE } \sigma \text{ AND } \mu \\ \text{ARE CONSTANTS} \end{array}$$

(B) Gaussian density function is often used as a dither source.

Fig. 3-8. Probability density functions.

one or one-half quantization increment is added to the audio signal. With this technique the added signal must be above audibility yet below the Nyquist frequency to prevent aliasing. Also, there is danger of modulation effects with the audio signal.

An analog dither signal necessarily decreases the S/N ratio of the digitization system; Blesser has proposed a system with a digital dither signal which would preserve the S/N ratio, as shown in Fig. 3-9. Noise is a random-valued signal and it may be simulated by generating a quickly changing pseudo-random sequence of digital data; this can be accomplished with a series of shift

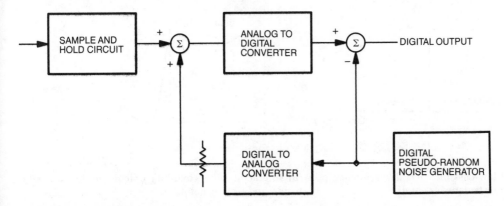

Fig. 3-9. Digital pseudo-random dither.

registers and a feedback network comprised of exclusive OR gates, and inputting that sequence into a D/A converter to produce analog noise. That signal may be added to the audio signal to achieve the benefit of dither, then following the signal A/D converter it may be digitally subtracted from the signal, leaving the dynamic range of the original signal. A further benefit is that inaccuracies in the A/D converter are similarly randomized.

3.3 Input Low-Pass Filter

The analog signal is low-pass filtered by a very sharp cutoff filter to band limit the signal and its entire harmonic content to frequencies below half the sampling frequency. For example, on a professional recorder with a sampling rate of 48 kHz, the filter's cutoff will be set around 22 kHz to allow for maximum attenuation at the half-sampling point. The input filter is sometimes called the anti-aliasing filter.

3.3-1 Filter Requirements • To prevent aliasing, the input low-pass filter must attenuate all frequencies above half the sampling frequency, yet not affect the lower frequencies. Thus an ideal filter would have a flat passband, an immediate, or "brickwall" filter characteristic, and an infinitely attenuated stopband, as shown in Fig. 3-10A. In addition to these frequency response criteria, an ideal filter would not affect the phase of the signal. Although in practice an ideal filter may be approximated, it presents a number of engineering challenges; to construct a brickwall cutoff means compromise on other specifications such as flat passband and low phase distortion. To alleviate the problems of a brickwall response, we could design filters with a more gradual cutoff because they do not exhibit such phase nonlinearities, however the frequency of the half sampling point would have to be increased to assure that it was placed in a sufficiently attenuated part of the filter characteristic. Thus a higher sampling frequency, perhaps three times higher than that required for a sharp cutoff filter, would be needed and storage requirements would be raised. To limit the sampling rate, and make full use of the band space below the half sampling point, a brickwall filter is the only alternative. It is ironic that this filter presents difficulties because it must be of analog design; after the A/D converter the same filtering could be accomplished digitally with little trouble.

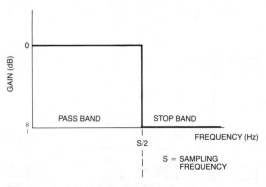

(A) An ideal low-pass filter has flat frequency response and instantaneous cutoff.

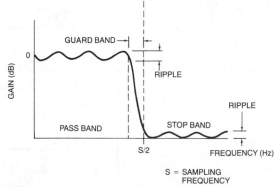

(B) In practice, filters exhibit ripple in the stop and pass bands, and sloping cutoff.

Fig. 3-10. Low-pass filter characteristics.

In professional recorders with a sampling frequency of 48 kHz, the input filters are usually designed for flat response from DC to 22 kHz; this provides a guard band of 2 kHz to ensure that attenuation is sufficient. The passband must have flat frequency response; in practice some frequency irregularity, called ripple, exists, but can be designed to be ±0.1 dB or less. The stopband's attenuation is designed to be equal or better than the system's dynamic range, as determined by the word length. A 16 bit system would require stopband attenuation of over 95 dB. A practical low-pass filter characteristic is shown in Fig. 3-10B.

Two other important filter criteria are ringing and phase linearity. Sharp cutoff filters exhibit resonance near the cutoff frequency and this ringing can cause coloration in frequency response, as shown in Fig. 3-11. The sharper the cutoff the greater the propensity to ringing. Certain filter types have inherently reduced ringing. Phase response is also a factor; analog tape machines, microphones and loudspeakers have always introduced phase distortion and low-pass filters also exhibit frequency dependent delays called group delays near the cutoff frequency, causing phase distortion. This can be corrected with an analog circuit preceding or following the filter, which introduces compensating delays to achieve overall linearity; pure delay, which is inaudible, results. In the cases of ringing and group delay, there is debate on the threshold of audibility of such effects. They occur at the extreme edge of hearing and it is unclear how perceptive the ear is to such high frequency phenomena.

Lagadec and Stockham have hypothesized that time domain dispersion might be responsible for audible distortion in sharp cutoff filters. In one experiment, a filter with ripple of less than ±0.5 dB and linear phase exhibited audible artifacts with a test signal. Analysis showed a pre-echo in the filter's

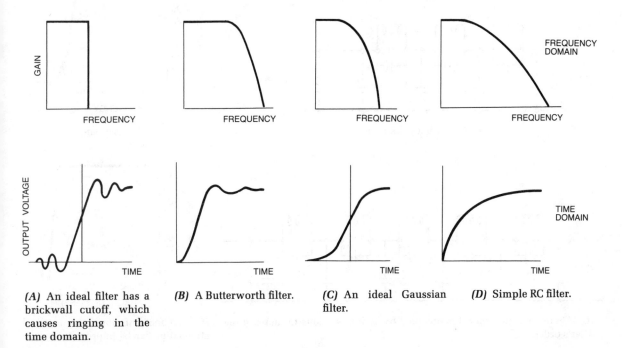

(A) An ideal filter has a brickwall cutoff, which causes ringing in the time domain.

(B) A Butterworth filter.

(C) An ideal Gaussian filter.

(D) Simple RC filter.

Fig. 3-11. Filter characteristics in frequency and time domain.

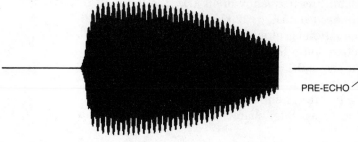

(A) Original 400 Hz burst test tone.

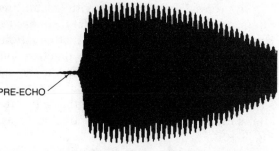
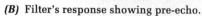

(B) Filter's response showing pre-echo.

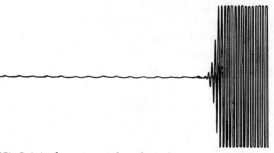

(C) Original tone's attack (enlarged).

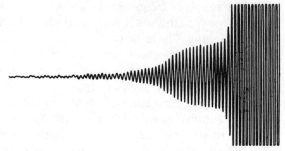

(D) Filter response's attack (enlarged) showing pre-echo.

Fig. 3-12. Experimental evidence of dispersion in sharp cutoff filter.

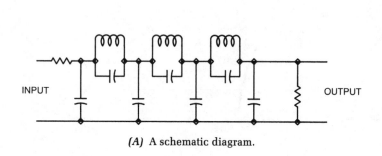

(A) A schematic diagram.

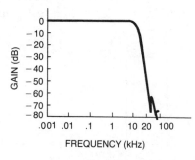

(B) Frequency characteristic.

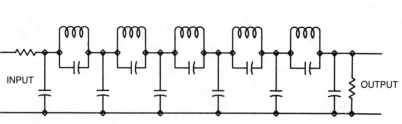

(C) Attenuation to −90 dB is obtained by adding sections to increase the filter's order.

(D) Steepness of slope and depth of attenuation can be improved.

Fig. 3-13. Passive low-pass Chebyshev filter.

output, as shown in Fig. 3-12, as well as a post-echo, both 32 dB below the amplitude of the test tone, and 40 milliseconds before and after it. These echoes were linked to the filter's ripple, and the researchers have speculated that a ripple specification of ± 0.001 dB might be required to place the echo pair below the threshold of audibility.

3.3-2 Filter Design • Given required filter specifications, several filter types may be employed, for example, corresponding to criteria incorporated in mathematical polynomials such as Bessel, Butterworth, or Chebyshev polynomials. For each of these, a basic design stage may be repeated, or cascaded to increase the filter's order, and to sharpen the cutoff. Thus, higher order filters more closely approximate the ideal filter's brickwall frequency response. For example, a passive Chebyshev low-pass filter is shown in Fig. 3-13; the results improve dramatically when the filter's order is increased through cascading. An active Chebyshev filter, with its frequency and phase response is shown in Fig. 3-14. The slope becomes steeper as "n," the filter order, is increased, as shown in Fig. 3-14B. Phase shift increases as n is increased, as shown in Fig. 3-14C. Commercial digital systems might employ a thirteenth order filter to achieve the required brickwall characteristic. For a given filter order, Che-

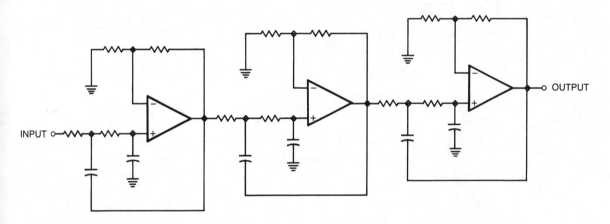

(A) Schematic diagram.

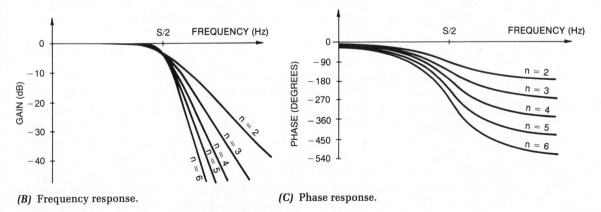

(B) Frequency response.

(C) Phase response.

Fig. 3-14. Sixth order Chebyshev filter using op amps.

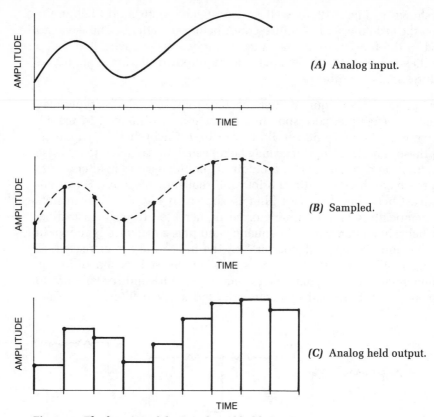

(A) Analog input.

(B) Sampled.

(C) Analog held output.

Fig. 3-15. The function of the sample and hold circuit.

byshev and elliptical low-pass filters give a closer approximation to the ideal than Bessel or Butterworth, but Chebyshev can show some ripple in the passband and elliptical can produce severe phase nonlinearities. Bessel can be made to approximate a pure delay, however its characteristic response requires higher orders. Butterworth is usually flat in the passband, but can exhibit slow transient response. Whichever design is chosen, and whatever order, the intent remains the same, to band limit the input audio frequency before it enters the digital system, while introducing no other anomalies. Analog filter design has always been challenging, with the advent of digital audio systems it is more so.

3.4 Sample and Hold

As its name implies, the sample and hold (S/H) circuit performs two simple yet crucial operations; it time samples the analog waveform at a fixed and periodic rate and holds the analog value of the sample until the analog-to-digital converter outputs the corresponding digital word. This is an important function because otherwise the analog value could change after the designated sample time thus the A/D could output incorrect digital words. The input and output response of a sample and hold circuit are shown in Fig. 3-15. The circuit is relatively simple to design; however, it must accomplish both of its tasks accurately; samples must be taken at precisely the correct time and the held value must not be allowed to vary.

3.4-1 Sample and Hold Requirements • As we have seen, time and amplitude information is used to completely characterize any acoustical waveform. The sample and hold circuit is responsible for helping to capture both informational aspects from the low-pass filtered analog waveform. Time information is stored implicitly, that is, time values are not stored separately, rather samples are taken at a fixed periodic rate, and reproduced at the same fixed periodic rate thus time information is preserved. The sample and hold circuit accomplishes this time sampling. A clock, an oscillator circuit which outputs timing pulses, is set to the desired sampling frequency and this signal is used to control the S/H circuit. An S/H circuit is essentially a capacitor and a switch; the circuit tracks the analog signal until the sample command causes the digital switch to isolate the capacitor from the signal, and the capacitor holds that analog voltage during the A/D conversion. An ideal sample and hold circuit is shown in Fig. 3-16. Varying sample times would result in errors; to prevent this the S/H circuit must be carefully designed and the sample command must be accurately clocked.

Variations in absolute timing, called jitter, as illustrated in Fig. 3-17, would create modulation noise; it would be particularly significant in the case of a high amplitude, high frequency input signal. Fortunately, jitter is unlikely in a

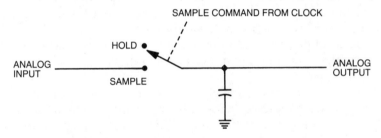

Fig. 3-16. **Ideal sample and hold circuit.**

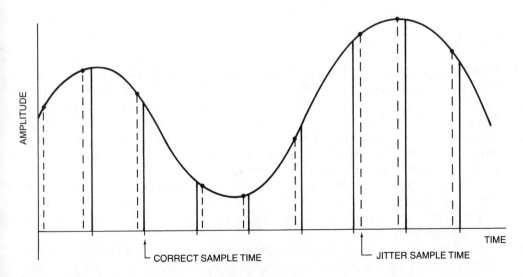

Fig. 3-17. **Time axis variations in sampling cause jitter.**

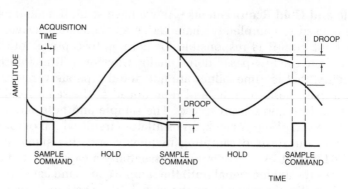

Fig. 3-18. **Two error conditions in the sample and hold circuit: acquisition time and droop.**

properly designed system. For example, timing is fundamentally controlled by a clock designed with a highly accurate crystal quartz oscillator. Any timing error anywhere in the digitization chain would indicate a severe malfunction. More likely, sample time errors would occur as a result in differences in the time it takes for the S/H circuit to respond to its sample command. The time lag for the S/H output to agree with the analog input signal is known as the acquisition time. This error causes a sampled value different from the one which occurred at the correct sample time. A fixed delay could be corrected; however, the delay is a function of the amplitude of the analog signal. The best solution is prevention; it is thus important to limit the acquisition timing error to a time less than a sampling period. In general, acquisition time should be less than 5 microseconds.

The S/H circuit's other primary function is to hold the captured analog voltage while conversion takes place. It is important for this voltage to remain constant, any variation greater than a quantization increment would result in an error on the A/D output. In practice, the held voltage is prone to droop because of current leakage. Care in circuit design and selection of components can limit droop to less than one-half a quantization increment over a 20 microsecond period. Acquisition error and droop are illustrated in Fig. 3-18. Acquisition time is the time between the initiation of the sample command, and the taking of the sample. Droop is the variation in hold voltage as the capacitor slowly leaks between samples.

3.4-2 Sample and Hold Circuit Design • The demands of fast acquisition time and low droop are sometimes in conflict in the design of a S/H circuit. For fast acquisition time, a small capacitor value would be better, permitting faster charging time in response to the hold command. For droop, however, a large valued capacitor is preferred because it is better able to retain the sample voltage at a constant level for a longer time. Circuit designers have found that capacitor values of approximately 1 nanofarad can satisfy both requirements. In addition, high quality capacitors made of polycarbonate, polyethylene, or Teflon dielectrics are specified; these materials can respond quickly, hold charge, and minimize dielectric absorption and hysteresis, phenomena which cause voltage variations.

In practice, a S/H circuit must contain more than a switch and a capacitor; active circuits such as operational amplifiers must buffer the circuit to condition the input and output signals, speed switching time, and help prevent

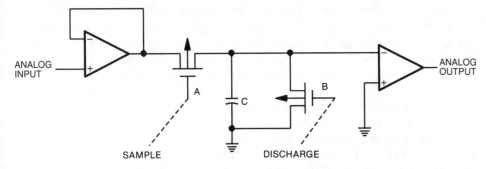

Fig. 3-19. **An example of a practical sample and hold circuit with JFET switches.**

leakage. Only a few specialized operational amplifiers will meet the required specifications of large bandwidth and fast settling time. Junction Field Effect Transistor (JFET) operational amplifiers are usually chosen as performing best. Thus, a complete S/H circuit might have a JFET input operational amplifier to prevent source loading and speed switching time, isolate the capacitor, and supply capacitor charging current. The S/H switch itself is probably a JFET device, chosen to operate cleanly and accurately with minimal jitter, and the capacitor will be high quality. A JFET operational amplifier is usually placed at the output to help preserve the capacitor's charge. An example of a practical sample and hold circuit is shown in Fig. 3-19. Switch A is closed to sample. After conversion, switch B is closed to discharge capacitor C and prepare for another sample.

The sample and hold circuit thus time samples and stores analog values for conversion. Its output signal is an intermediate signal, a discrete staircase of the original analog signal, but still not a digital word.

3.5 Analog-to-Digital Conversion

The analog-to-digital converter lies at the heart of the recording side of an audio digitization system, and it is the most critical and costly component in the entire electronic system. This circuit must determine which quantization increment is closest to the analog waveform's current value, and output a binary number specifying that level, accomplishing that task in 20 microseconds or less. Fortunately, several types of circuits are available for this operation; two fundamental analog-to-digital design approaches prevail. The input analog voltage can be compared to a variable reference voltage within a feedback loop to determine the output digital word, or the input voltage can be allowed to decrease and the time it takes to reach zero is timed with a counter which generates an output digital word. Successive approximation and parallel methods are examples of the former, integration methods are examples of the latter.

3.5-1 Analog-to-Digital Converter Requirements • The A/D converter must perform a complete conversion at each sample time, for example, 48,000 conversions per second per channel in a professional audio digitization system. Furthermore, the digital word it provides must be an accurate representation of the input binary voltage. In a 16 bit linear PCM system each of the 65,536 increments must be evenly spaced throughout the amplitude range so that even the least significant bits in the resulting word are meaningful. Thus, speed and accuracy are key requirements for any A/D converter.

The time it takes for an A/D converter to output each digital word is called its conversion time. For an A/D converter conversion time must be within the span of one sampling period. It is sometimes difficult to achieve accurate conversion from sample to sample because of settling time or propagation time errors; the result of accomplishing one conversion may influence the next. If a converter's input moves from voltage A to B, and then later from C to B, the resulting digital output for B may be different because of the device's inability to properly settle in preparation for the next measurement. Obviously, dynamic errors grow more severe with demand for higher conversion speed. In practice, speeds required for full fidelity audio digitization can be achieved, indeed some A/D converters simultaneously process two waveforms, alternating between left and right channels, however cost is always relatively high for any A/D with fast conversion time.

Accuracy is another important consideration; several specifications have been devised to evaluate the performance of A/Ds. Integral linearity measures the "straightness" of an A/D's output; it describes how close the transition voltage points, the analog input voltages at which the digital output changes from one code to the next, are to a straight line drawn through them. Integral linearity is illustrated in Fig. 3-20. Linearity is tested and the reference line is drawn across the converter's full output range. Integral linearity is one of the most important A/D specifications, An "n" bit converter is not a true "n" bit converter unless it guarantees at least ± ½ LSB linearity. The converter in Fig. 3-20 has a ± ¼ LSB integral linearity.

Differential linearity error is a measure of the distance between transition voltages, that is, the widths of input voltage bands. Differential linearity is illustrated in Fig. 3-21. Ideally, all of the bands of an A/D transfer function should be 1 LSB wide. A maximum differential linearity error of ± ½ LSB means that the input voltage may have to increase or decrease as little as ½ LSB or as much as 1½ LSB before an output transition occurs. If this specification was exceeded, to perhaps ± 1 LSB, some levels could be 2 LSBs wide and others would be 0 LSB wide, that is, that output code would not exist. The

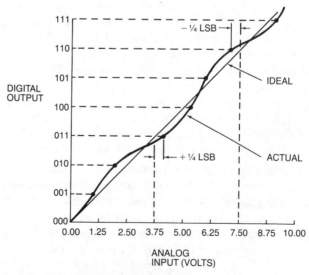

Fig. 3-20. **Integral linearity specification of an A/D converter.**

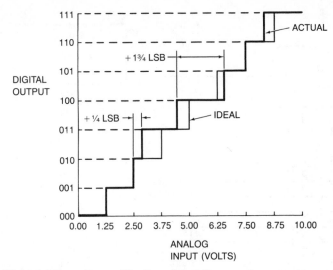

Fig. 3-21. **Differential linearity specification of an A/D converter.**

converter in Fig. 3-21 has an error of $\pm \frac{3}{4}$ LSB; some levels are $+ \frac{1}{4}$ LSB wide, others are $+ 1\frac{3}{4}$ LSB wide.

Absolute accuracy error, shown in Fig. 3-22, is the difference between the supposed level at which a digital transition occurs, and where it actually occurs; a good A/D should have an error of less than $\pm \frac{1}{2}$ LSB. An offset voltage, gain error or noise error can all affect this specification. For the converter in Fig. 3-22, each interval is $+ \frac{1}{8}$ LSB in error. In practice, otherwise good A/D devices can sometimes drift with temperature variations and thus introduce inaccuracies.

3.5-2 Successive Approximation Analog-to-Digital Converter • There are many types of A/D circuit designs, appropriate for various applications. For audio digitization, the necessity for both speed and accuracy limits the choices to a

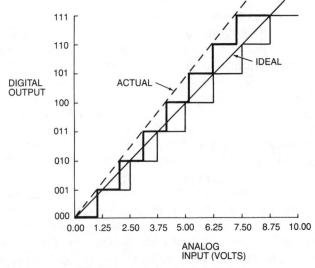

Fig. 3-22. **Absolute accuracy specification of an A/D converter.**

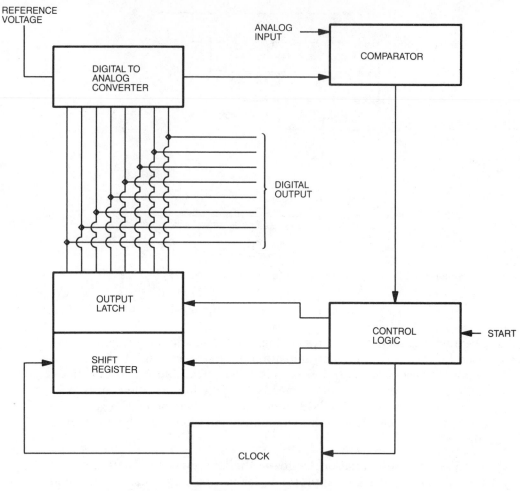

Fig. 3-23. **Block diagram of a successive approximation register A/D converter.**

few types. The classic A/D used in audio digitization is the successive approximation register (SAR) A/D design. It is shown in the block diagram in Fig. 3-23. This converter employs a digital-to-analog converter in a feedback loop, a comparator, and a control section. In essence, this converter compares the analog input with its interim digital word converted into analog, until the two agree within the given resolution. In operation, the device follows an algorithm which bit by bit sets the output digital word to match the analog input.

For example, let's assume an analog input of 6.92 volts, and an 8 bit A/D converter; the operational steps of the SAR converter is shown in Fig. 3-24. The most significant bit in the SAR is set to 1, with the other bits still at zero, thus the word 10000000 is applied to the internal D/A. This word places the D/A output at its half value of 5 volts. Since the input analog voltage is greater than the D/A output, the comparator remains high; bit one is stored at logical one. The next most significant bit is set to 1, and the word 11000000 is applied to the D/A, and an interim output of 7.5 volts appears. This is too high so the second bit is reset to zero and stored. The third bit is set to 1, and the word 10100000 is applied to the D/A, which produces 6.25 volts, so the third bit

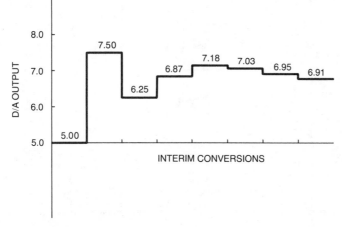

SUCCESSIVE APPROXIMATIONS:

10000000	= 5.00000 V	OK
11000000	= 7.50000	RESET
10100000	= 6.25000	OK
10110000	= 6.87500	OK
10111000	= 7.18750	RESET
10110100	= 7.03125	RESET
10110010	= 6.95312	RESET
10110001	= 6.91406	OK

Fig. 3-24. **Intermediate steps in an SAR conversion.**

remains high. This process continues until the least significant bit is stored and the digital word 10110001 representing a converted 6.92 volts is output from the A/D. This successive approximation method requires "n" D/A conversions for every one A/D conversion, where "n" is the number of bits in the output word. In spite of this recursion, SAR converters offer relatively high conversion speed and are very cost effective because they can convert "n" bits for the cost of one bit's worth of converter. The SAR has thus gained wide acceptance in audio digitization systems.

3.5-3 Dual Slope Integrating Analog-to-Digital Converter • Another method of analog-to-digital conversion which is increasingly being used in audio circuits is dual slope conversion. In a single slope converter, as shown in the block diagram of Fig. 3-25, a reference value is allowed to integrate as a ramp function, and the time it takes to equal the input value is measured by a high-speed counter; the count is output as the digital word representing the analog value. For a 16 bit conversion in 10 microseconds, the clock would have to run at a rate of 6 gigahertz, which is too fast for conventional digital circuits.

In a dual slope design, the integrating method is modified by using two counters; both are timed to yield 8 bits each. The clock would now have to run at 50 megahertz, which is an attainable frequency. In a dual slope converter, as shown in Fig. 3-26, two current sources can be switched to the integrator. The coarse ramp and the fine ramp, are both timed by digital counters, which output the data word. The first unity current is switched to form the coarse ramp and this is timed with the first counter; one counting period equals 256 periods of the second counter. Then $\frac{1}{256}$ of the first current is switched to form the fine ramp which is similarly timed by the second counter. The counting

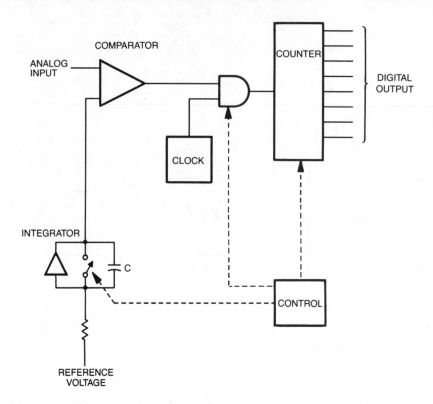

Fig. 3-25. **Single slope integrating A/D converter.**

slopes are shown in Fig. 3-27. The most and least significant 8 bits are joined to form a 16 bit digital output word. Linearity of a dual slope A/D circuit is largely dependent on the dielectric loss of the capacitor. This method has been successfully integrated onto a single chip and is being used in several digital audio systems.

A successive approximation register A/D thus uses a D/A converter in a loop; it tries a digital word, converts it and compares the analog result to the original input, then corrects its approximation until the proper digital word has been determined and output. Integrating A/D converters use a timing circuit; a capacitor stores a voltage then a timer counts as the voltage is integrated as a ramp. The number of counts in that timing becomes the output digital word. Whichever method is used, the goal of digitizing the analog signal has been accomplished.

For digitization systems in which real-time processing such as delay and reverberation is the aim, the signal is ready for processing through software or dedicated hardware. In the case of a digital recording system, further processing is required to prepare the data for the storage medium.

3.6 Record Processing

After the analog signal has been converted into a binary number, several operations must take place prior to storage on the medium. Although specific

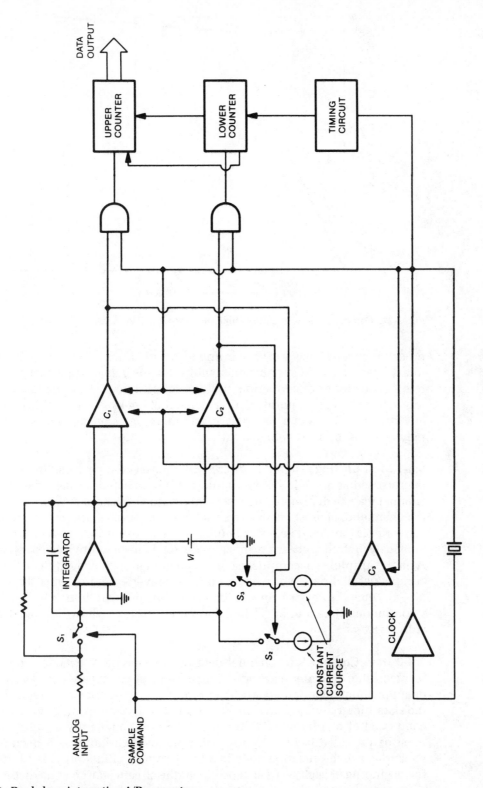

Fig. 3-26. Dual slope integrating A/D converter.

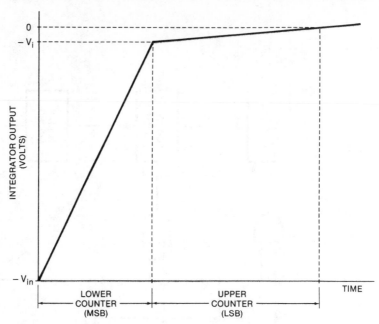

Fig. 3-27. **Counting slopes in a dual slope integrating A/D converter.**

processing techniques vary according to the medium to be used (e.g., fixed or rotary head), all PCM systems must multiplex the data, add inspection bits and redundancy for error correction, perform interleaving, and provide format coding. While there is an uninteresting element of bookkeeping in this processing, any drudgery is crucial in preparing the data for storage, and ensuring that playback will be accomplished satisfactorily.

3.6-1 Multiplexing • Digital audio recording is a serial process, that is, the data is processed as a single stream of information. In actual storage, data from one channel may be distributed over several recorded tracks, but before that occurs, the channel data must begin as a serial stream. However, the output of the A/D converter is parallel data, for example, entire words of 16 bits are output simultaneously. Before processing, this parallel data must be converted to serial data. A data multiplexer accomplishes this transformation; the multiple input circuit accepts parallel data words from the A/D converter and outputs the data one bit at a time. Successive conversions of 16 bits each are thus transformed into a continuous stream of bits. The 16 input multiplexer shown in Fig. 3-28 uses shift registers.

3.6-2 Data Coding • Raw channel data must be properly encoded to facilitate its recording, and later recovery. Several kinds of coding are applied to modify or supplement the original data. An example of an encoded bit stream showing the data hierarchy is shown in Fig. 3-29. The time-multiplexed data code is comprised of one frame after another; to prevent ambiguity a coding scheme must be provided to identify the beginning of each frame as it occurs in the stream. A synchronization code is a fixed pattern of bits provided to identify the beginning of each word as it occurs in the stream chosen to be distinct from any naturally occurring data bit pattern, in much the same way that a comma

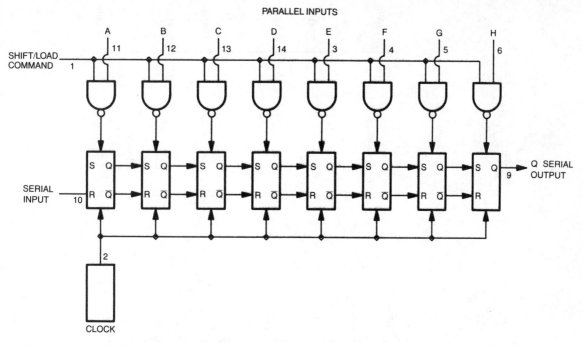

(A) Functional block diagram of 74165 shift register.

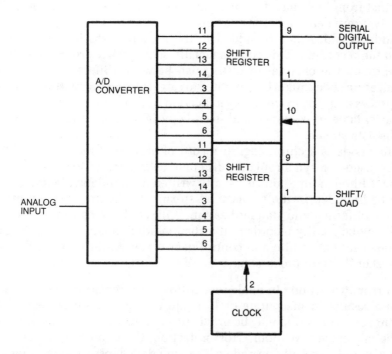

(B) Conversion from parallel to serial with cascaded 74165 shift registers.

Fig. 3-28. **Parallel load shift registers can be used to convert the parallel output of the A/D converter to serial data.**

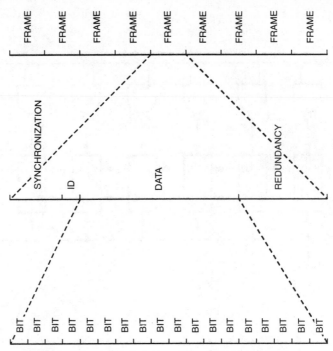

Fig. 3-29. Data encoding. Prior to digital recording, data is encoded according to a specific hierarchy of frame format.

is distinct from the characters in a sentence. The potential error of inability to correctly group data frames is alleviated.

Address codes may be added to the bit stream to identify locations of data in the recording. This code is usually sequentially ordered and is distributed through the channel to distinguish between different sample sections. Identification codes might be generated to carry information pertinent to the playback processing. For example, specification of sampling frequency, use of pre emphasis, table of contents, and even copyright information may be entered into the data stream.

Time code is a chronological record of absolute time in a recording and may be used for synchronizing different recorders, or for use in electronic editing. Although time code is usually written to an independent track, it could easily be included in the primary bit stream. The entire question of codes is thus one of format and standardization. With analog recording, only an audio signal may be readily recorded on a channel; digital audio offers the opportunity to consolidate audio and control data, a step toward greater flexibility in audio recording and processing.

3.6-3 Error Protection • In analog recording, an error occurring in the storage medium results in inaccurate or lost information. In digital recording error detection and correction may be provided such that the effect of storage defects may be minimized. Without error protection, the quality of digital audio recording would be greatly degraded. The high data density required for efficient storage necessitates a system able to correct for storage errors upon playback. Several steps are taken for error protection; the data is processed before storage by adding parity bits and check codes, both of which are redundant data created from the original data to help detect and correct errors. To complete the error protection scheme, to prevent a single defect from destroying both the original data and that required to correct for its loss, interleaving is employed in which data is scattered to various locations on the medium. An example of interleaving is shown in Fig. 3-30. A complete discussion of parity, check codes, redundancy, interleaving, and the entire error protection system is presented in Chapter 6.

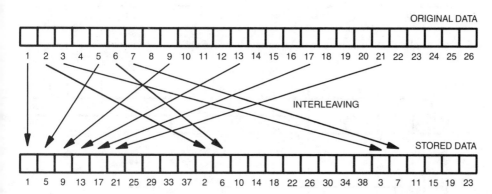

Fig. 3-30. **An example of interleaving.**

3.6-4 Format • The transformation process from raw data to coded data is dependent on the selection of format. Each of these record processing steps may be carried out in many ways, and the way in which the resulting data is assembled may be variously determined. A digital recording may consist of many frames of data, arbitrarily interleaved. Each frame consists of group codes, such as synchronization, address, identification, data, and redundancy. Each frame contains many data words, including samples which contain the time-multiplexed bits of audio data. There is obviously considerable latitude involved in determination of a format, and the relative efficiency and success of each are not equal; additionally the choice of medium strongly influences format design. This is discussed in further detail in Chapter 5.

3.7 Modulation Processing

Modulation processing is the final electronic manipulation of the audio data before its storage. Because digital audio is commonly considered to involve the storage of 1s and 0s, it may be surprising to learn that the binary code is not recorded directly, rather a modulated code is stored, which represents the bit stream. It is thus a modulation waveform which is recorded and interpreted upon playback to recover the original binary data, and thus the audio waveform. Modulation codes have been aptly described by Stockham as the handwriting of digital audio.

3.7-1 Necessity for Modulation • Digital media storage creates a number of specific difficulties which can only be overcome through modulation techniques. The words in a binary bit stream are marked with a synchronization code to distinguish between them, but there is no way to directly distinguish between the individual bits. A series of 1s or 0s, as shown in Fig. 3-31, would form a static signal upon playback; if no other timing or decoding information is available, time information implicitly encoded in the bit periods would be lost. It is necessary to record the data in such a way that pulse timing is delineated. Additionally, it is inefficient to store binary code directly on the

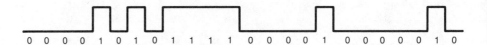

0 0 0 0 1 0 1 0 1 1 1 1 0 0 0 0 1 0 0 0 0 0 1 0

Fig. 3-31. **Modulation is needed to differentiate between strings of consecutive 1s or 0s.**

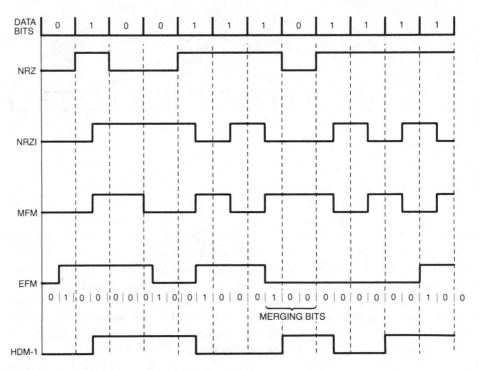

Fig. 3-32. **Examples of modulation codes.**

medium. Much greater densities with high data code fidelity may be achieved through methods in which modulation code fidelity is quite low. Pulse code modulation is not suitable for recording directly in light of these requirements thus other modulation techniques must be devised specifically for recording purposes. While binary recording is thus concerned with storing the 1s and 0s of the data stream, the signal actually recorded may be quite different; typically, it is transitions from one level to another rather than the amplitude levels themselves which represent the information on the medium.

3.7-2 Modulation Codes • For reasons stated previously, all binary recording systems employ modulation for more efficient data storage. Their common method of using only two values gives them the inherent advantage of digital storage; relatively large variations in the medium will not affect data recovery.

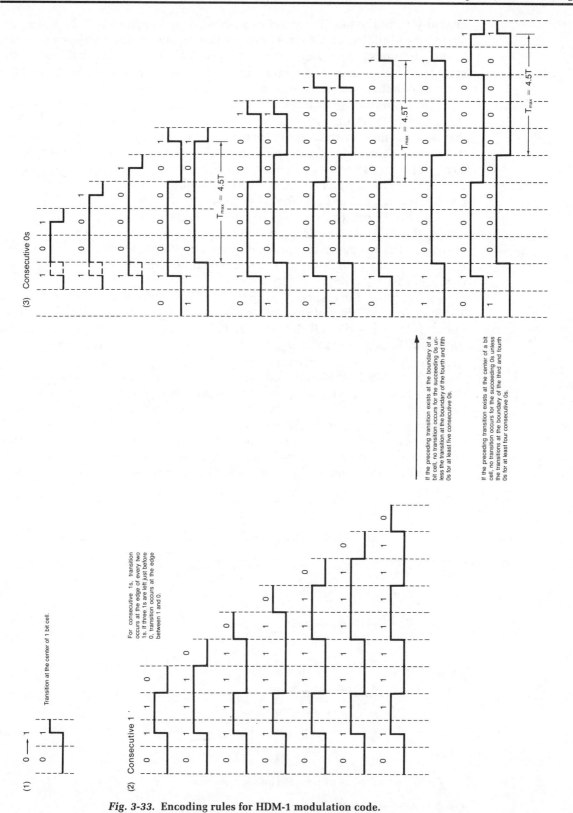

Fig. 3-33. Encoding rules for HDM-1 modulation code.

Digitally stored data is robust, thus high packing densities may be achieved. Various modulation codes have been devised to encode binary data according to the medium's properties; of many, only a few are applicable to digital audio storage, on either magnetic or optical mediums. Several of the codes applicable for audio digitization are shown in Fig. 3-32.

The Nonreturn to Zero (NRZ) code is the oldest and most basic form of modulation; 1s and 0s are represented directly as high and low levels. The direction of the transition at the beginning or end of a bit cell indicates a 1 or 0. It suffers from one of the problems which encourages use of modulation; strings of 1s or 0s cannot be directly synchronized. Thus, NRZ is used only for digital audio recording methods in which synchronization is externally generated. The video cassette recorder is an example of a medium using NRZ.

The Nonreturn to Zero Inverted (NRZI) code is similar to the NRZ code, except that only 1s are denoted with amplitude transitions. Thus, any flux change in the magnetic medium indicates a 1. Transitions occur only in the middle of a bit cell. NRZI is used as an intermediate modulation code in Compact Disc encoding and decoding.

In Modified Frequency Modulation (MFM) code, sometimes known as Miller code, the code performs either a positive or negative-going transition, in the middle of the bit cell, for each 1. There is no transition for 0s, rather a transition is performed at the end of a bit cell only if a string of 0s occurs. Synchronization can be accomplished with MFM.

Eight-to-Fourteen Modulation (EFM) code is used to store data on Compact Discs; it is an efficient and highly structured code. Blocks of 8 data bits are translated into blocks of 14 bits using a look-up table, a kind of dictionary, which assigns an arbitrary and unambiguous word. The 1s in the output word are separated by at least two 0s so that the minimum distance between transitions is three channel bits. Each logical 1 represents a transition in the medium; for example, on a Compact Disc each logical 1 is physically present as a pit

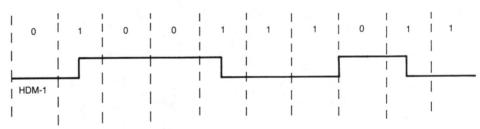

(A) The channel bits of the modulated waveform.

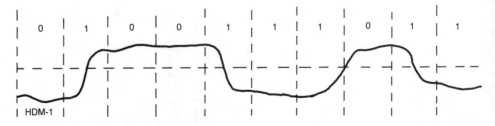

(B) The signal as it might be recorded on magnetic tape.

Fig. 3-34. **Digital Magnetic Storage.**

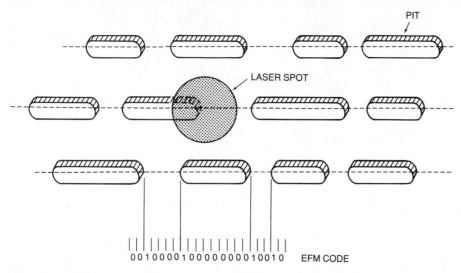

Fig. 3-35. **Digital Optical Storage. Data is encoded as a series of pits on the recording surface of a Compact Disc.**

edge. Greater recording density is facilitated; with EFM a 25 percent gain in data density is achieved over unmodulated coding.

The HDM-1 modulation code is used in professional digital recorders constructed with the Digital Audio Stationary Head (DASH) format. Its encoding rules, shown in Fig. 3-33, illustrate the amount of signal data which may be compactly stored in the modulation code. The minimum wavelength is 50 percent longer than conventional codes such as MFM thus greater recording density is permitted. The HDM-1 code was devised by Doi and other engineers.

3.7-3 Recording • Following modulation the data is ready for storage on the medium. In the case of a fixed head digital recorder, the data is applied to a recording circuit which generates the current necessary for saturation recording. The flux reversals recorded on the tape thus represent the bit transitions of the modulated data. The recorded waveforms, as shown in Fig. 3-34, illustrate the dirty nature of the signal on the tape; this does not affect the integrity of the data, and permits the recording of higher densities. In optical systems such as the Compact Disc, a previously recorded digital tape is played through a laser cutting machine which produces the master glass plate used in CD manufacturing. The EFM code results in pits, each pit edge represents a binary 1, while spaces between represent binary 0s, as shown in Fig. 3-35. These and other topics are considered in Chapters 5 and 7.

Chapter 4

Digital Audio Reproduction

Introduction

In an audio digitization system, the recording and reproduction processors serve as input and output transducers, converting the analog audio signal into a signal suitable for digital storage, then reconverting the stored or processed signal to analog form. In a linear pulse code modulation (PCM) digitization system the function of the subsystems of the reproduction side of the signal path are largely reversed from the function of the record side. The reproduction subsystems include the demodulation circuit, reproduction processing circuits, demultiplexer, digital-to-analog converter, output sample and hold circuit, and output low-pass filter. This chapter describes the reproduction circuits used in a linear PCM audio digitization system, as shown in Fig. 4-1.

4.1 Demodulation Circuits

The demodulation circuits are the first step in the reproduction of the digital audio signal, the system in which the coded waveform recorded on the tape is again converted to an analog signal. The demodulation circuits must accom-

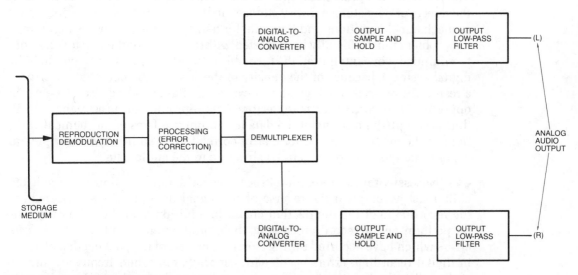

Fig. 4-1. **Linear PCM reproduction section.**

plish several important functions. The signal derived from the reproduce head is of very low amplitude thus it must be amplified. This waveform is very distorted and must be processed to recover the data. Finally, the data must be synchronized and demodulated to restore the original literal data.

4.1-1 Operation of Demodulation Circuits • A preamplifier is required to boost the signal from the tape head. The signal is so low in level that any processing can be accomplished only after amplification. To achieve high recording density, the fidelity of the modulation code waveform as recorded on the medium was allowed to deteriorate. Thus, the signal from the tape does not have the clean characteristics of the original data. Rather, the amplitudes of the recorded data as read from the tape head are rounded and only the transitions between levels correspond to the original signal. A waveform shaper circuit is used to identify the transitions and thus reconstruct the 1s and 0s of the signals. Data can be entirely recovered thus there is no penalty for the waveform's deterioration. The data is again as clean as if it had been literally recorded, but storage of a much greater amount of data has been permitted.

The music signal data and its error code is identified and separated from the peripheral data which is additionally identified and separated into frame synchronization and bit synchronization signals. Frame synchronization pulses are used to identify individual frames and bit synchronization pulses are derived and used to identify individual bits within each frame and synchronize the playback signal and thus determine the 1 or 0 content of each pulse. The modulated music signal data, whether it is HDM-1, EFM, MFM, or another code, is demodulated to NRZ code, that is, a simple code in which amplitude level represents the binary information. The method for interpreting NRZ is to read a logical 1 when there is a high amplitude and logical 0 when there is a low level amplitude. The music data has thus regained its original binary form and is ready for further reproduction processing.

4.2 Reproduction Processing

The reproduction processing circuits are primarily concerned with minimizing the effects of data storage. Every storage medium suffers from limitations, such as mechanical variations and potential for damage to data. With analog storage, the problem must generally be corrected within the medium itself; for example, to minimize wow and flutter, the turntable's speed must be kept accurate. With digital systems, because of the density of the storage, the potential for error as a result of storage is much greater; however, digital encoding also presents the opportunity to correct for many errors. The reproduction processing circuits thus accomplish buffering of the data to minimize effects of mechanical variations in the medium, and perform error correction. In addition, demultiplexing is performed to restore the parallel structure to the audio data.

4.2-1 Necessity for Reproduction Processing • The reproduction processing circuits must accomplish the reverse of the signal manipulations performed in the record side of the digitization chain. In addition, and the primary reason for performing the record processing, the reproduction circuits must check for errors which have occurred during storage. Because of the packing density used in digital recording, errors occur with certainty, only their frequency and severity will vary. If within tolerance, errors may be detected and corrected with absolute fidelity to the original data thus making digital recording a virtually lossless technique. If the nature of the errors exceeds the error correction circuit's ability to correct, estimated values may be substituted for missing or bad data.

Transport problems include rapid variation in timing, low frequency drifting of timing, errors from tape stretching, and improper head alignment. A badly designed digital recorder could additionally cause crosstalk errors in the heads or associated wiring, or errors due to power supplies with electrical noise. Additionally, errors are caused by both manual and electrical editing; reproduction processing circuits must be able to accommodate the large disruptions of data which occur at an edit point.

4.2-2 Description of Reproduction Processing Circuits • The reproduction processing circuits must initially de-interleave the data. Prior to recording, the data was scattered in the bit stream to ensure that a defect in the medium would not affect too many consecutive bits. With de-interleaving, the data is again properly assembled, and bit errors caused by medium defects are now scattered through the bit stream, where they are easier to correct because of their isolation. The entire interleave and de-interleave process is shown in Fig. 4-2.

Mechanical instability in the medium transport will introduce timing errors, such as jitter, as data is read from the medium—this is shown in Fig. 4-3; to overcome this problem a data buffer is used. A buffer may be thought of as a pail of water; water is poured in carelessly, but the spigot at the bottom of the pail supplies a steady stream of water. Specifically, a buffer is a memory into

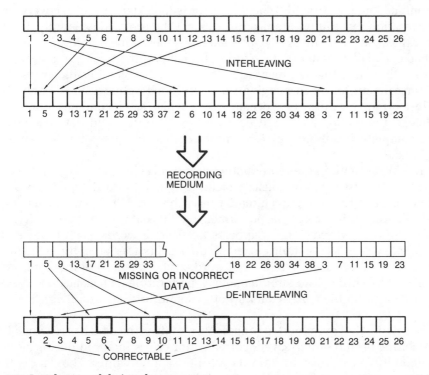

Fig. 4-2. **Interleave and de-interleave process.**

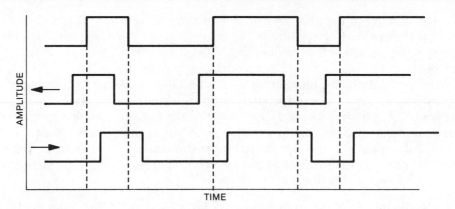

Fig. 4-3. Jitter.

which the data is fed, irregularly, as it arrives from the medium. However, the output of the buffer occurs at an accurately controlled rate thus ensuring precise data timing. Samples are thus assembled at the same rate at which they were taken, guaranteeing the lossless nature of time sampling.

Using redundancy techniques, such as parity and checksums, the data is checked for errors. When the parity bits or checksums which are calculated do not agree with those read from the medium, an error has occurred either in the audio data, or the parity and checksum data. Several methods are used to isolate the error and determine where the fault has occurred. In the case of bad audio data, error correction techniques are used to recover the correct values. Using parity bits, checksums, or redundant data, the missing values may be determined and substituted. When the error is too extensive for recovery, error compensation techniques are used to conceal the error. Most simply, the last data value can be held until valid data resumes. Linear interpolation is a method of calculating new data to form a bridge over the error. For larger errors, interpolation and other compensation techniques become insufficient, and error concealment becomes marginal; the presumed values differ widely from the lost original values. In extreme cases, when error compensation is not sufficient, the audio signal will be switched off until valid data resumes. A more complete discussion of error protection techniques may be found in Chapter 6.

The final circuit in the reproduction processing chain is the demultiplexer. The serial bit stream now consists of the original audio data, or at least as original as the error protection circuitry has achieved. However, one remaining manipulation which must be performed on the data is to convert it to its parallel form in which it again appears as discrete words, each representing one sample value. The demultiplexer circuit accepts a serial bit input, counting as the bits are clocked in. When a full word has been received, it outputs all of the bits of the entire word simultaneously, performing its task again and again as the data is applied. An example of a demultiplexer circuit is shown in Fig. 4-4.

Following the reproduction processing circuitry, the data has regained timing stability, been de-interleaved, corrected for errors incurred during storage, and demultiplexed to again form its parallel sample words. The data is then ready for digital-to-analog conversion.

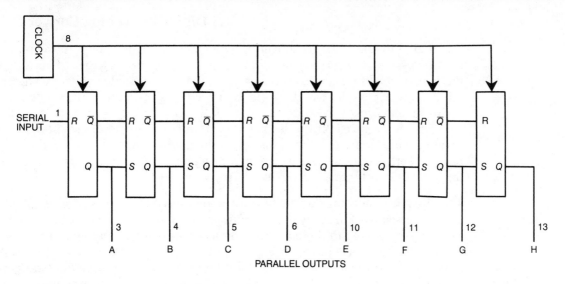

(A) Functional block diagram of 74164 shift register.

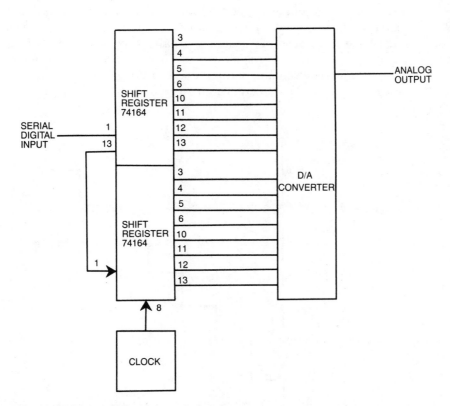

(B) Conversion from serial to parallel with cascaded shift registers.

Fig. 4-4. **Parallel output shift registers can be used to convert the serial output of the reproduction circuitry to parallel data prior to D/A conversion.**

4.3 Digital-to-Analog Conversion

The digital-to-analog (D/A) converter is the most critical element in the reproduction system; just as the analog-to-digital (A/D) converter largely determines the overall quality of the record system, the digital-to-analog converter determines how accurately the digitized signal will be restored to the analog domain. In playback-only systems, such as the Compact Disc system, the D/A converter must be carefully designed to permit stable operation under many varying conditions, especially as encountered by automobile and portable players. Fortunately, several excellently designed D/As are available, and those integrated circuits are available at relatively low cost.

4.3-1 Requirements for the Digital-to-Analog Converter • The digital-to-analog converter is subject to many of the requirements and prone to many of the same errors as the analog-to-digital converter, as described in Chapter 3. The ideal transfer function, as shown in Fig. 4-5, is more nearly approximated by D/A converters. A D/A converter must exhibit integral linearity, that is, the "straightness" of its transfer from digital to analog must be good. Its differential linearity must maintain an error of less than $\pm\frac{1}{2}$ LSB, that is, when the input code changes by one bit, the analog output should change by one voltage step. A D/A must be monotonic, that is, the analog output must always increase as the digital input is increased, and decrease as the input decreases. An example of a D/A converter which is not monotonic, that is, the output does not always increase as the digital input increases, as shown in Fig. 4-6. A D/A converter must have good absolute accuracy, small offset, and fast settling time. The criteria for setting time are shown in Fig. 4-7. Settling time for a D/A is the elapsed time between a new input code and the time when the analog output falls within a specified tolerance. In terms of audio performance specifications,

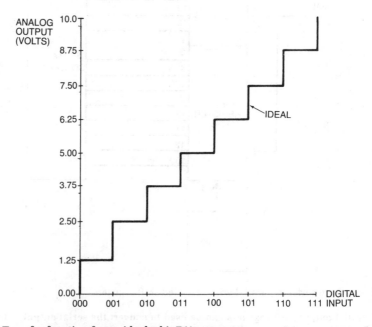

Fig. 4-5. **Transfer function for an ideal 3 bit D/A converter.**

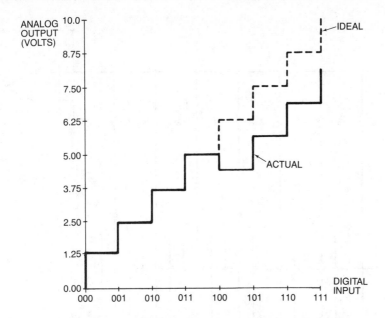

Fig. 4-6. Transfer function for a 3 bit D/A converter which is not monotonic.

these combined requirements constitute a device with low distortion and intermodulation products.

4.3-2 Weighted Resistor Digital-to-Analog Converter • Many types of digital-to-analog converters are available. Three types are commonly employed in audio digitization systems. To understand their function, we must begin with a simple design which illustrates the operation of the D/A. A digital-to-analog converter accepts an input digital word and converts it to an output analog voltage or current. The simplest kind of D/A contains a series of resistors and switches; it is known as a weighted resistor D/A converter, an example is shown in Fig. 4-8. There is a switch for each input bit and the corresponding resistor represents the value associated with that bit. A reference voltage is used to generate

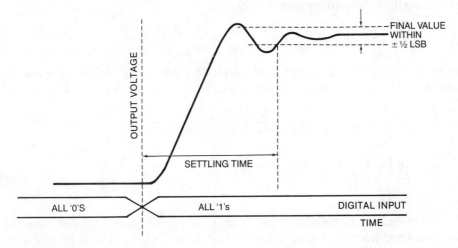

Fig. 4-7. Settling time for a D/A.

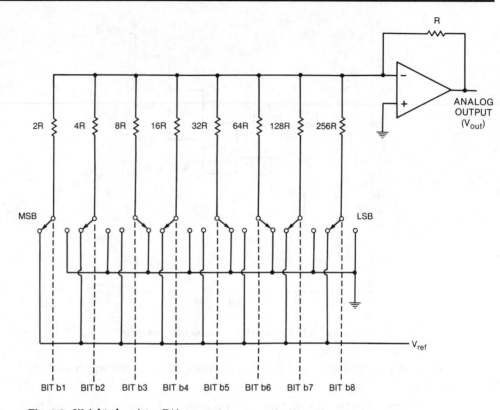

Fig. 4-8. **Weighted resistor D/A converter.**

currents in the resistors. A digital 1 closes a switch and contributes a current while a digital 0 causes the switch to remain open and no current flows. An operational amplifier sums the currents and converts them to an output voltage. A low value binary word with many 0s would keep many switches open and a small voltage would result, a high value word with many 1s would close more switches and a high voltage would result. While this design looks good on paper, it is rarely used in practice because of the complexity in manufacturing resistors with sufficient accuracy.

$$V_{out} = -V_{ref}\left(\frac{b1}{2} + \frac{b2}{4} + \frac{b3}{8} + \frac{b4}{16} + \frac{b5}{32} + \frac{b6}{64} + \frac{b7}{128} + \frac{b8}{256}\right)$$

where b1 through b8 represent the input binary bits. For example, suppose the reference voltage is 1, and the input word is 11010011, and $V_{ref} = 10v$,

$$V_{out} = -10\left(\frac{1}{2} + \frac{1}{4} + \frac{0}{8} + \frac{1}{16} + \frac{0}{32} + \frac{0}{64} + \frac{1}{128} + \frac{1}{256}\right)$$

$$= -10\left(\frac{1}{2} + \frac{1}{4} + \frac{1}{16} + \frac{1}{128} + \frac{1}{256}\right)$$

$$= -8.24v$$

Since each next resistor value must be a power of two greater than the previous one, widely varying values result, a condition difficult to manufacture.

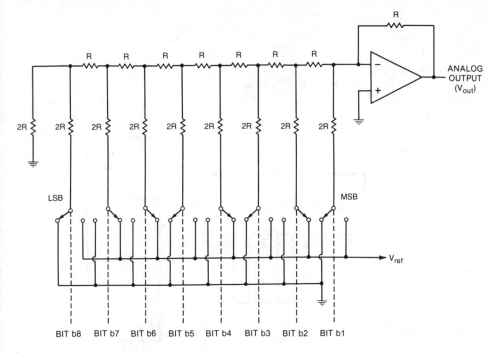

Fig. 4-9. **R-2R D/A converter.**

4.3-3 R-2R Ladder Digital-to-Analog Converter • A more suitable design approach for a D/A converter is the R-2R resistor ladder shown in Fig. 4-9. This circuit contains resistors and switches; however, there are two resistors per bit. Each switch contributes its appropriately weighted component to the output; the current splits at each node of the ladder resulting in currents through the switch resistors which are weighted by binary powers of two. Digital input bits are used to control ladder switches to produce an analog output

$$V_{out} = -V_{ref}\left(\frac{b1}{2} + \frac{b2}{4} + \frac{b3}{8} + \frac{b4}{16} + \frac{b5}{32} + \frac{b6}{64} + \frac{b7}{128} + \frac{b8}{256}\right)$$

With an input word of 01010110 and V_{ref} = 10v,

$$V_{out} = -10\left(\frac{0}{2} + \frac{1}{4} + \frac{0}{8} + \frac{1}{16} + \frac{0}{32} + \frac{1}{64} + \frac{1}{128} + \frac{0}{256}\right)$$

$$= -10\left(\frac{1}{4} + \frac{1}{16} + \frac{1}{64} + \frac{1}{128}\right)$$

$$= -3.36v$$

If a current I flows from the reference voltage, I/2 flows through the first switch, I/4 through the second switch, I/8 through the third switch, etc. The R-2R network is preferred because of ease of manufacture, only two values of resistors are needed. Precision values are needed only for the more critical bits. Stability with respect to temperature can be achieved with a compensation feedback loop. A high precision signal is generated and compared to the signal generated by the D/A. The difference between the two is applied to a memory which in turn outputs a correction word to the input of the D/A. Errors caused

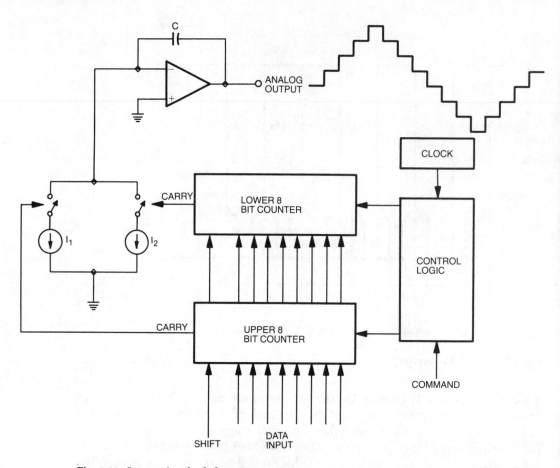

Fig. 4-10. **Integrating dual slope D/A converter.**

by variations in the components are thus self-corrected and distortion is minimized.

4.3-4 Integrating Digital-to-Analog Converter • A second approach to the design of the digital-to-analog converter is the dual slope integrating converter, as shown in Fig. 4-10. Its operation is similar to that of the integrating A/D converter. The inherent accuracy of conventional ladder D/As depends on the precision of the resistor values. Even a small error in the value of the most significant bit or second most significant bit resistors can greatly effect the accuracy of the output voltage. Integration techniques do not require a ladder, and if internal reference currents are carefully regulated, high accuracy is obtained. A dual slope integrating D/A converter can accept a 16 bit input word, dividing the word into the most significant, and least significant, 8 bits. An output integrating amplifier accepts the two current sources for timed intervals and outputs an analog voltage proportional to the timing ramps. This D/A design is easily produced on an integrated circuit, and is currently used in several digitization systems.

4.3-5 Dynamic Element Matching Digital-to-Analog Converter • A dynamic element matching (DEM) D/A converter as demonstrated by Van de Plassche and Dijkmans uses current dividing to perform its task. A series of switches and current sources are arranged in a cascade configuration; the input data word is used to open and close the switches, one switch per bit, and thus access a current source. Each source is a power of two less than the first unity source, thus each switch/source pair represents one bit of the input word. Accuracy of conversion depends on the precision of the divided current sources. In the dynamic element matching system, the exact factor of two ratio is generated by rapidly interchanging two currents which are divided by a factor of two from an input current source, thus any differences are averaged out; this is illustrated in Fig. 4-11. Other types of DEM converters use division by four to further reduce the magnitude of current error.

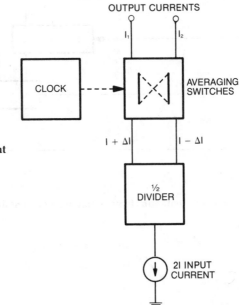

Fig. 4-11. **Dynamic element matching current divider stage.**

Individual currents for each switch/source pair are obtained by a cascade of current dividers, as shown in Fig. 4-12. The currents for the least significant bits may be derived by passive division, the other currents use active switching methods, with RC filters to smooth any ripple produced by switching. A complete DEM block diagram is shown in Fig. 4-13. This 14 bit converter accepts clocked serial input data and uses a reference current source and a cascade of dividers to produce the analog output current. A DEM converter may be fabricated on an integrated circuit without being overly complex, and its performance specifications are very good. Linearity of ±½ LSB between −20°C to +70°C may be obtained. Because of its fast conversion speed, this type of D/A may be used within A/D circuit designs.

A wide variety of D/A designs are now available and each manufacturer has chosen the design it feels is best. This competitive spirit greatly decreased the problems of manufacturing D/A (and A/D) converters, which had delayed the availability of digital processing for consumer audio applications. Still to come are further cost reductions as these and other circuit elements in the digitization system are consolidated into one or two integrated circuits, and

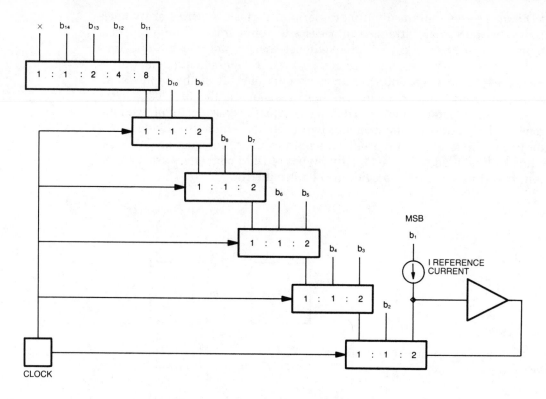

Fig. 4-12. Dynamic element matching current divider cascade.

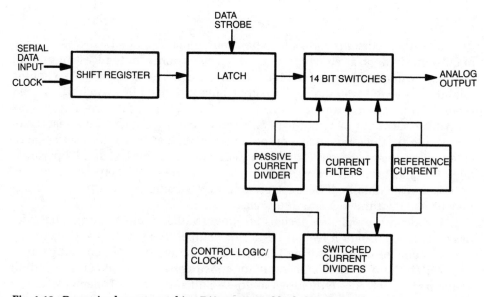

Fig. 4-13. Dynamic element matching D/A converter block diagram.

mass produced by audio manufacturers. This in turn will bring about further reductions in the overall cost of digital audio systems.

4.4 Output Sample and Hold Circuit

Perhaps surprisingly, most audio digitization systems require two sample and hold (S/H) circuits; one at the input to maintain the sampled analog value while the A/D converter performs its task, and another S/H circuit on the output to sample and hold the signal from the D/A converter, primarily to remove irregular signals called switching glitches. The output S/H circuit is sometimes called the aperture circuit.

4.4-1 Necessity for Output Sample and Hold • Most digital-to-analog converters can generate several kinds of erroneous signals which are superimposed on the analog output voltage. Digital data input to the D/A converter require a certain amount of time to stabilize before all of the bits have found their proper binary levels. During that period, the D/A's output voltage may reflect that uncertainty in the form of voltage variations. A carefully designed D/A can minimize this problem by latching the input data, that is, data is temporarily held before it is presented for conversion, to allow data values to settle. However, virtually all D/As suffer from another cause of erroneous data; when a D/A switches from one output voltage level to another, voltage variations known as switching glitches usually occur. Even D/A circuits with fast settling times can exhibit momentary glitches. If these false voltages were allowed to proceed to the digitization system's output they would be manifested as distortion.

To eliminate this possibility an output S/H circuit is used. The output S/H circuit acquires a voltage from the D/A converter only when that circuit has reached a stable output condition. That correct voltage is held by the S/H circuit during the intervals when the D/A converter switches between samples. Accuracy of data to the system output is assured. The operation of an output sample and hold circuit is shown in Fig. 4-14.

4.4-2 Requirements for Output Sample and Hold • From a general hardware standpoint the output sample and hold circuit can be designed similarly to that used at the input of the digitization system. An example of a sample and hold circuit is shown in Fig. 4-15. In some specifications, such as droop, the output S/H circuit may be less precise; any droop would result in only a DC shift at the digitization system's output, and this can be easily removed. In other respects the output S/H circuit must be carefully designed and implemented in a digital audio system. Because of its differing utility, the output sample and hold circuit often requires special attention to specifications such as hold time, and transition from sample to sample.

One overriding concern in the design of a S/H circuit is aperture error, an attenuation of high frequencies which is a natural consequence of the output staircase waveform. As Stockham has pointed out, the output of an ideal D/A converter would be simply impulses, each occupying an area equal to the original sample point; there would be no high frequency attenuation. Since an ideal D/A cannot be built, a S/H circuit is required, as previously described. Mathematically, the staircase waveform equals the ideal impulse output followed by a filter with an impulse response equal to a sample pulse (the stair-

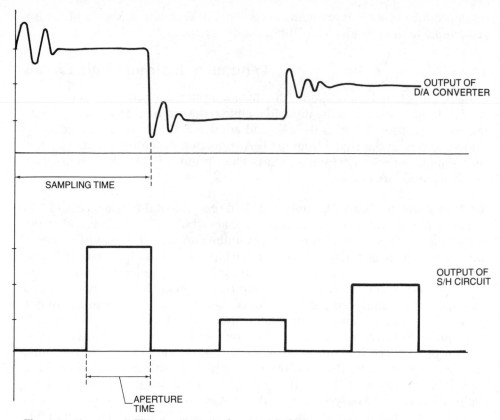

Fig. 4-14. An output S/H circuit is used to avoid glitches present at the output of a D/A converter.

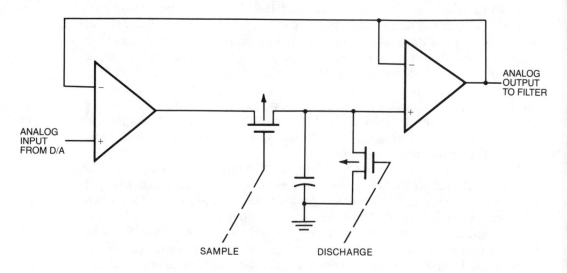

Fig. 4-15. Closed loop sample and hold circuit.

case is a convolution of the original samples by a square pulse of width of one sampling period). The resulting frequency response is the Fourier transform of the sample pulse. The spectrum of a staircase of pulses naturally attenuates at high frequencies. This differs from the original flat response thus a frequency response error called aperture error results, in which high frequencies might be attenuated by as much as 4 dB. (The frequency response of the Fourier transform has the form sin (x)/(x) which at half sampling frequency equals 0.64, which is about 4 dB.) Fortunately, this effect can be controlled; the error can be minimized by approximating the output of the impulse train of the ideal D/A converter by manipulating the duration of the hold time in the sample and hold circuit. If the hold time is decreased relative to the duration of the sample time, the attenuation effect can be made almost negligible, as shown in Fig. 4-16. If hold time is set to one-quarter of a sample period, the amplitude decrease at 20 kHz is only 0.2 dB. This is considered optimal because a shorter hold time would begin to lessen the signal to noise ratio of the system. An elegant solution is digital compensation of the data prior to D/A conversion; however, this requires considerable computation. Another technique used in overcoming aperture error is to design a frequency compensation into the low-pass filter which succeeds the S/H circuit. Alternatively, a pre-emphasis high frequency boost could be applied at the system input, then the naturally occurring deemphasis at the output S/H would again result in a flat response.

An output sample and hold circuit is primarily used to eliminate errors caused by the D/A converter during transition, however the S/H must function so as to avoid introducing transition errors of its own. A S/H circuit outputs a steady value while in the hold mode; however, when switching to the sample (or acquisition) mode a slow transition would introduce incorrect intermediate values into the staircase voltage. This problem was extraneous in the input S/H because the A/D converter accomplished its digitization during the hold mode and ignored the transition mode. However, the output S/H is always connected to the system output and any transition error would appear at the output. To overcome this problem, the output S/H circuit must switch as quickly as possible from hold to sample mode. While in theory this would greatly minimize the possibility of distortion caused by the transitions, in practice it is difficult to achieve the necessary high slew rate, which has been

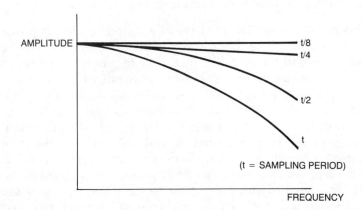

Fig. 4-16. Aperture error. As aperture time is decreased, the effect is minimized.

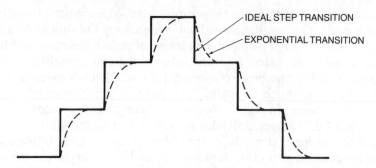

Fig. 4-17. **An integrate and hold circuit would produce an exponential release from step to step.**

calculated to be as high as 5000 V/microsecond by Blesser. Thus, an additional modification to the basic S/H circuit is often carried out. An exponential change in amplitude from one quantization interval to the next does not create nonlinearity in the signal. An integrate and hold output is shown in Fig. 4-17. It can be shown that an exponential transition from sample to sample causes only a slight high frequency deemphasis at the output. A S/H circuit which integrates the difference between its current and next value results in an exponential transition, thus an integrating S/H circuit is often used. The attenuation of high frequencies in an integrate and hold circuit is much less than that already produced by the sample and hold process itself.

The output S/H circuit thus removes switching glitches from the D/A converter's output voltage. To avoid introducing its own distortion, an integrating function is typically used, and hold time is set to less than a sample period to minimize aperture error. Following the output S/H circuit, the staircase analog signal is ready for filtering, the final step in the de-digitization process.

4.5 Output Low-Pass Filter

The first, and last, circuits in a complete audio digitization system are low-pass filters, known as the anti-aliasing and anti-imaging filters, respectively. Although their designs can be almost identical, their function is always very different. In addition to the classic analog anti-imaging filter, new digital filter designs using oversampling techniques have been developed.

4.5-1 Necessity for Output Filters • Given the criteria of the Nyquist sampling theorem, the function of the input low-pass filter is clear; it must filter out all frequencies above half the sampling frequency to prevent aliasing. The necessity for using a low-pass filter at the output of the digitization system to similarly filter out all frequencies above half the sampling frequency stems from the need to convert the output pulse amplitude modulation (PAM) signal to a continuous waveform. An examination of the signal following the output of the S/H circuit, as in Fig. 4-18, shows the staircase nature of the analog waveform, a characteristic which is an artifact of the sampling process. The sudden shifts in level represent high frequency components not present in the original signal. Clearly something must be done to transform the stair-steps into a smooth continuous waveform. An output low-pass filter performs this function by filtering out the high frequencies, leaving the original waveform, as shown in Fig. 4-19. In other words, the staircase signal is smoothed; the output filter is, in fact, sometimes called a smoothing filter.

Viewing the process from a more mathematical standpoint, we can observe how the situation arose as a result of sampling. Specifically, we multiplied the time domain audio signal with the time domain sampling (pulse) signal. In terms of the spectrum of these two sampled signals, this convolution produced a new sampled spectrum identical to the original unsampled spectrum; however, the spectra are infinitely repeated across the frequency domain at multiples of the sample rate, as shown in Fig. 4-20. For example, an original 1 kHz sine wave sampled at 44 kHz would also create components at 43, 45, 87, and 89 kHz, etc. Although the sample and hold process substantially reduces the amplitude of the extraneous frequency bands, significant components still remain after the S/H circuit, particularly in the region near the audio band as shown in Fig. 4-21. To convert the original sampled information back into correct analog information, we must destroy all of the look-alike spectrums, leaving only the frequency-correct spectrum. This is accomplished by low-pass filtering.

Sharp-eyed readers might question the need to filter out frequencies such as 43 or 89 kHz, since they lie so far above the threshold of human audibility.

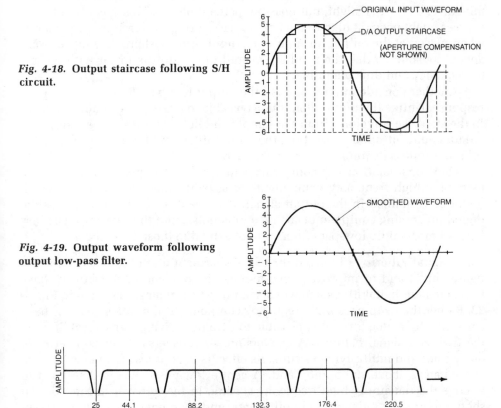

Fig. 4-18. Output staircase following S/H circuit.

Fig. 4-19. Output waveform following output low-pass filter.

Fig. 4-20. The frequency spectrum produced by sampling showing the periodic repetitions of the audio band.

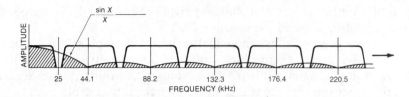

Fig. 4-21. **The effect of sample and hold. The remaining signal components are shown shaded.**

The original waveform is reproduced without filtering but the accompanying spectra could cause modulation in other downstream equipment through which the signal passes, that in turn could negatively affect the audio signal. Other digital systems might be immune because their input filters would remove the high frequencies, but oscillators in analog recorders or transmitters could conceivably create difference frequencies in the audible band.

4.5-2 Requirements of Output Filters • The design criteria for the output low-pass filter are similar to those of the input filter; the pass band should be flat, and the stop band highly attenuated, for sampling efficiency the cut-off should be sharp. Audibility of phase shifts caused by sharp-cut filters must be considered with output filters; as with sharp-cut anti-aliasing filters, group delay occurs. The threshold of audibility of phase distortion has not been determined. The effect is slight, but possibly perceptible. Phase shifts of up to 360 degrees might occur at high frequencies. Phase correction circuits are available, when placed at the output (or input) of a digitization system, they make group delays constant so that the resulting overall phase shift is linear with respect to frequency and thus inaudible.

Another consideration unique to the output low-pass filter is its transient response. Unlike the input filter, it must be able to process the sudden changes in the staircase waveform from the sample and hold circuit. Just as a slow S/H circuit could introduce distortion, the output filter could create unwanted by-products unless its transient response is good.

One consideration not commonly addressed is the possible presence of extremely high frequency components of several hundred megahertz which might be contained in the output signal. Because of its high speed operation, digital processing equipment can create this noise, and the filtering characteristics of most audio low-pass filters do not extend to those frequencies.

4.5-3 Digital Filtering • To avoid the use of sharp-cut output filters, to decrease phase shifts, and to improve signal-to-noise ratio, some manufacturers have implemented digital filters using oversampling techniques, as shown in Fig. 4-22. Rather than suppress high frequency components after the signal has been converted to analog form, it is possible to employ a digital transversal filter on the digitized signal before D/A conversion. A transversal filter is a series of delay lines and multiplying circuits; its effect is to suppress frequency spectra immediately above the audio band. A gentle low-pass filter is then used to remove the remaining high frequency spectra. In the transversal filter design shown in Fig. 4-23, delays and multipliers, and a summer, are used to accomplish the oversampling filtering. Each delay is equal to one sampling period, and the multipliers use 12 bit coefficients, a different coefficient for each multiplier. Each 16 bit data word is multiplied four times with different coefficients

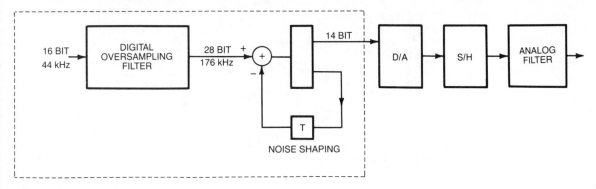

Fig. 4-22. **Digital filter with oversampling.**

before it is passed to the next delay. The product of each multiplier is 28 bits (16 + 12). When these products are summed, a weighted average of a large number of samples is obtained. The result is the multiplication of the sampling frequency, and a cutoff filter characteristic. For example, if the sampling frequency is multiplied by four, the oversampler would extend 44.1 kHz to 176.4 kHz. Four times as many samples are present after multiplication. The overall effect is the suppression of the frequency bands between 20 kHz and 156.4 kHz. As can be seen in Fig. 4-24, the bands centered at 44.1, 88.2, and 132.3 kHz have been removed. In this example, the multipliers' coefficients create a sloping filter characteristic between 20 kHz and 24.3 kHz. An important benefit

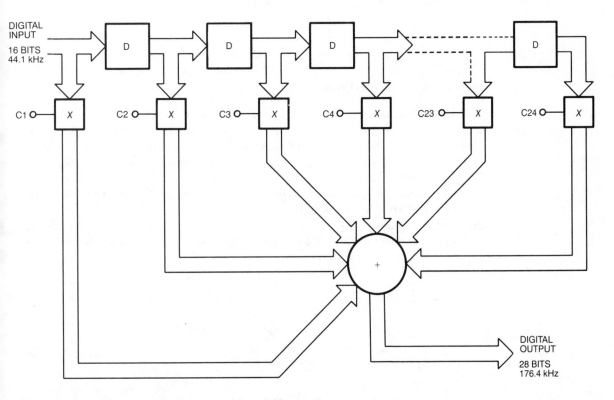

Fig. 4-23. **Twenty-four element digital transversal filter.**

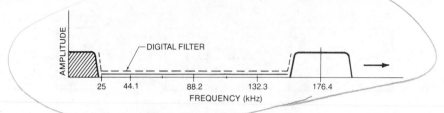

Fig. 4-24. **Oversampling multiplies the sampling frequency.**

of oversampling is a decrease of quantization noise. For example, when a signal is sampled with a four times higher sampling frequency, the noise density is reduced to one-fourth in the audio band, yielding a 6 dB improvement in the S/N ratio. In addition, high frequency attenuation due to aperture error can be compensated for in the transversal filter by choosing coefficients to create a slightly rising characteristic prior to cutoff.

A noise shaping circuit can be employed to further minimize audio band noise and thus increase S/N ratio, as shown in Fig. 4-25. The 28 bit data words from the transversal filter are rounded off to create most significant 14 bit words.

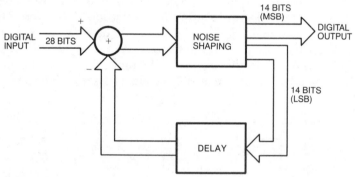

Fig. 4-25. **Noise shaping circuit.**

The 14 least significant bits are delayed by one sampling period and subtracted from the next data word. The result is a 7 dB improvement in S/N ratio in the audio band. At frequencies approaching 88.2 kHz, this feedback slowly becomes in phase with the input and noise increases, as shown in Fig. 4-26. However, this is filtered out by the S/H circuit and analog low-pass filter.

Oversampling permits the noise shaper to transfer the information in the fifteenth and sixteenth bits of the 16 bit signal into a 14 bit output with duty cycle modulation of the fourteenth bit. This permits the use of a 14 bit D/A converter to process the 16 bit signal. Full 16 bit information capability is

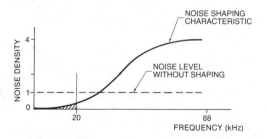

Fig. 4-26. **Noise shaping suppresses noise in the lower frequencies.**

retained because the information capacity of the two "surplus" bits is transferred into the four times higher information transmission level of the oversampled signal. In other words, the averaged value of the quarter-length 14 bit samples is as accurate as that of 16 bit samples over the original sample period; this is illustrated in Fig. 4-27.

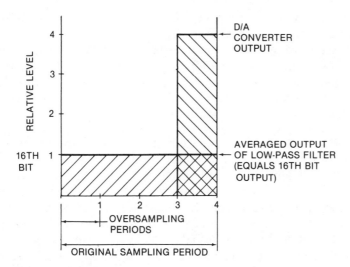

Fig. 4-27. Sixteen bit accuracy is retained in 14 bit oversampling.

A S/H circuit processes four samples of the finely grained step waveform within every 44.1 kHz sampling period. The sin(x)/(x) response produces a null of 176.4 kHz, and the sidebands surrounding 176.4 kHz are attenuated, as shown in Fig. 4-28. Finally, a gently sloping analog low-pass filter, with little phase shift, is used to remove the remaining components in the band around 176.4 kHz, leaving the original audio band. For example, a third-order Bessel design with a −3 dB point at 30 kHz might be employed. A primary advantage to digital filtering is avoidance of large phase shifts which characterize brickwall filters, especially at high frequencies. A digital filter might introduce a ±0.5 degree angle of deviation. As we have seen, further advantage of oversampling and noise shaping is the extension of S/N ratio. Oversampling gives a gain in S/N ratio of 6 dB and the noise shaper improves the ratio by a further 7 dB. Thus, a 14 bit converter with oversampling and noise shaping will yield S/N performance comparable to that of a 16 bit D/A converter and brickwall filter. Digital filter techniques have applications in both consumer and professional products.

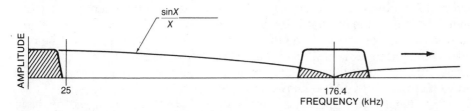

Fig. 4-28. Oversampling eliminates the lower frequency bands; the hold function leaves only the shaded component.

The output low-pass filter is followed by an analog audio amplifier to restore the signal to line levels, ready for power amplification. With this last subsystem, we have arrived at a complete PCM digital audio system design.

4.6 A Complete PCM System

Following our discussions of the recording and reproduction sections of a PCM digitization system, it might be useful to review and summarize the hardware subsections which comprise this audio system. In addition, no system is complete without an analysis of its performance specifications. In the case of digital audio systems, new specifications still need to be devised to properly evaluate these systems' performance. A block diagram of the recording and reproduction sections of a PCM system is shown in Fig. 4-29.

4.6-1 Recording Section • In classic stereo PCM digitization the audio signal is sampled, quantized, converted to binary form, and encoded for recording or transmission. Reversing the process produces a replica of the original signal. The recording section consists of input amplifiers, a dither generator, input low-pass filters, sample and hold circuits, analog-to-digital converters, a multiplexer, digital processing and modulation circuits, and, of course, a storage medium such as digital tape. Alternately, as in the case of digital delay lines, digital reverberators, and other real-time signal processors, the recording medium would be replaced by active digital processing circuitry. The design of the input and output digitization chain would remain essentially unchanged; however, there would be no need for modulation and demodulation, and error protection circuitry.

Aside from the requirement of absolute quality, so as not to compromise the fidelity of the ensuing digital system, there is nothing unique in the design of the input amplifier to a digital audio system. The dither generator is a controlled noise circuit typically outputting white noise. The analog signal is low-pass filtered by a very sharp cutoff filter to band limit the signal and its entire harmonic content to frequencies below half the sampling frequency. On a professional recorder with a sampling rate of 48 kHz, the filter cutoff will be set around 22 kHz to allow for maximum attenuation at the half-sampling point. A number of analog filter designs may be employed for this purpose, such as types corresponding to Bessel, Butterworth, or Chebyshev polynomials. All of these designs offer a flat pass band, sharp cutoff and a low stop band.

The input sampler samples discrete values of the input signal at a fixed periodic rate, and it holds the analog value while the analog-to-digital conversion takes place. This is required because a varying input to the A/D circuit could result in error. A sample and hold circuit is essentially a capacitor and a switch. The circuit tracks the signal until the sample commands causes the switch to isolate the capacitor from the signal; the capacitor holds that analog voltage during conversion. The timing of the sample command must be carefully regulated to prevent jitter, the phenomenon of incorrectly varying sample times. Furthermore, the capacitor must be carefully chosen and isolated to prevent any loss of voltage, known as droop.

The signal now appears as a staircase, a hybrid analog signal ready for conversion. The analog-to-digital converter is the most critical and costly component in a digitization system; consider that this circuit must transform the

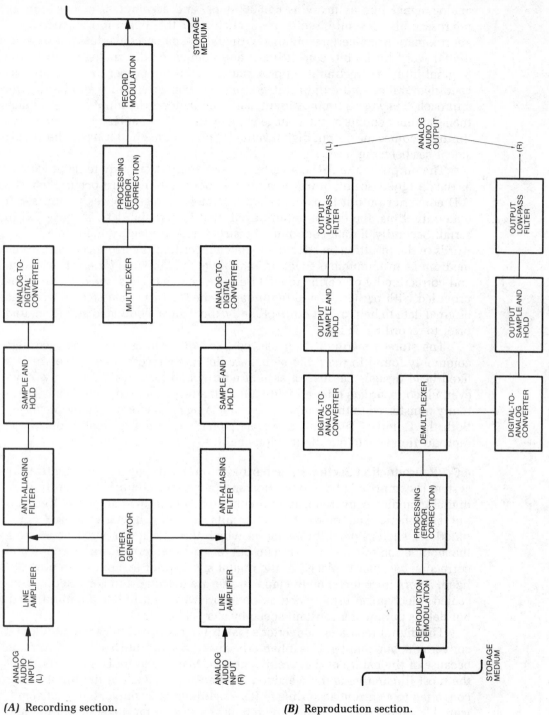

(A) Recording section.　　　　　　　　　*(B)* Reproduction section.

Fig. 4-29. A complete PCM digitization system.

analog signal into as many as 65,536 steps, and accomplish that task in 20 microseconds. Several circuits are available for this operation. A successive approximation converter contains a digital-to-analog converter; it forms its digital word bit by bit, converting it and comparing the analog result to the original input, correcting its approximation until the proper digital word has been determined and output. Integrating A/D converters offer another design approach. They use a timing circuit; a capacitor stores the input analog voltage then the timer counts as the voltage is integrated. The number of counts in that timing becomes the output digital value. Whichever method is used, the analog signal has been digitized.

The output of the A/D converter is raw binary data ready to be processed as the designers see fit; however, a number of operations must occur. First, the A/D converter output is parallel data and most storage devices permit serial data only; thus, the data is multiplexed, that is, parallel data is converted to serial. Secondly, a data code must be structured to identify the original data words in the resulting bit stream. In analog recording, an error occurring in the medium is an immutable event. In digital we may provide for error detection and correction by preparing the data prior to storage. Thus, the data stream is provided with parity bits and redundancy checks, extra data created from the original data to help detect errors. Finally, the data is modulated and formatted prior to recording on tape.

The storage medium itself can be fixed or rotating head tape recorders commonly found in the professional studio, many recording engineers prefer fixed head designs because of ease of editing and historical familiarity, however, video recorders offer a highly efficient storage method. The data is eventually transferred to the Compact Disc or analog media for consumer playback. Both the Compact Disc and the professional's recorder generally follow the reproduction side of our digitization chain.

4.6-2 Reproduction Section • The hardware on the output side of a digitization system is comprised of demodulation and processing circuits, a demultiplexer, digital-to-analog converters, aperture circuits, output low-pass filters, and output amplifiers. The reproduction circuits accomplish many housekeeping functions, such as demultiplexing, in which the data steam is recovered from the modulation scheme, and is again put into parallel form. Error detection and correction safeguards placed in the digital signal prior to storage are now utilized. Errors introduced by the tape or disc are detected and corrected or concealed. Mechanical errors such as transport wow and flutter are eliminated; our data is output at a constant speed by a crystal clock.

The digital-to-analog converter's task is the reverse of the analog-to-digital converter's, but simpler. It is inherently easier to accomplish a D/A conversion because of the nature of the circuit's design. Many types of D/As are used but the most illustrative is the weighted resistance D/A. Each of the input bits is converted to a current according to its weighting; for example, the most significant bit would yield a higher value than the second most significant bit, and so on, with each bit value changing twofold. The currents are added and converted to a voltage which corresponds to the original voltage prior to the A/D conversion. Other D/A converter designs include those using dual slope integrating methods. For economy's sake, sometimes one D/A converter is shared between channels.

An output sample and hold circuit essentially consists of a switch which is timed to wait for the D/A conversion. When the D/A output voltage is stable, the switch passes the voltage. This removes unstable values and corrects for aperture error of the pulse amplitude modulation (PAM) signal present at this point. It is gated as a function of the original sampling frequency.

The output low-pass filter is substantially identical to the input low-pass filter and performs the same cutoff function. The staircase function is smoothed and all of the high frequency components of those transitions are thus removed and the original waveform is recovered. Either an analog filter may be used, or a digital filter (oversampling) technique, in which the sampling frequency is multiplied and a more gentle cutoff filter may be employed. With digital filtering, the oversampling precedes the D/A converter in the signal chain. The final part of our digitization system is the analog amplifier, again hopefully designed with care (by analog designers, not digital ones).

4.6-3 PCM System Criteria • Analog recordings have limited frequency response, limited dynamic range, noise, distortion, wow and flutter, and generation loss. Similarly, digital recordings have limited frequency response, limited dynamic range, noise, distortion, wow and flutter, and generation loss, but in different amounts and different connotations. Any comparison of the circuits comprising digital and analog systems illustrates the lack of engineering similarities. Although both systems may enable the high fidelity recording of audio, their techniques are so different that it is difficult to meaningfully compare their performance by employing the same specifications. Traditional specifications measure an analog system's quality, but perhaps they do not entirely apply to a digital system. New, digital specifications are needed.

Even with the simplest measurements such as harmonic distortion, it is difficult to compare an analog performance specification to a digital one. For example, in an analog system, distortion increases with increased signal level, whereas with digital it decreases with increased level, thus a specification of 0.003 percent total harmonic distortion (measured at 0 dB, or maximum recording level) for a digital system will conceal much higher distortion measurements at lower amplitude levels. On the other hand, increasing distortion with decreasing level might not be a serious consequence. For example, consider a hypothetical case of increasing distortion in an undithered system which measures 0.003 percent total harmonic distortion (THD) at 0 dB, 0.03 percent at −40 dB, and 0.3 percent at −80 dB. While 0.3 percent appears high, it represents a distortion level 80 dB below the 0 dB reference level thus the overall effect of the distortion will be quite small.

As another example, phase response of digital audio systems is rarely fully specified; the phase of a system might begin to deviate sharply at 5 kHz, and exceed 360 degrees of phase shift at 20 kHz in a system with brickwall filters. The audibility of high frequency phase shift is still not known, but that further demonstrates the need for meaningful specification. The effect of phase shift on imaging and localization must be fully understood. Similarly, in systems using one D/A converter for two channels, unless corrected one channel is always delayed relative to the other—is this audible, and under what conditions, and to what extent? Clearly, as digital audio systems uncover new subtleties in our hearing ability, specifications will need revision.

For now, we can examine the basic performance specifications of a digital audio recorder or reproducer. As with any tape machine or disc player, dynamic range, distortion, frequency characteristics, and wow and flutter may be measured. Dynamic range and distortion in a digital audio system are primarily dependent on the quantization level of the A/D, and the linearity of the A/D and D/A converters; dynamic range of over 90 dB, and less than 0.005 percent total harmonic distortion at 1 kHz and 0 dB, and intermodulation distortion less than 0.005 percent at 0 dB. Frequency characteristics are dependent on sampling frequency and the low-pass filters; a response from 5 to 20 kHz ±0.5 dB is common. Wow and flutter is a function only of the accuracy of crystal clocking; this should be negligible. Channel separation at 1 kHz should be more than 90 dB.

An entirely new performance specification is required to properly evaluate digital audio system's error protection. We require ways of measuring how robust the recorded data is, and how forgiving the equipment is to dust, dirt, and defects at playback time. Several manufacturers produce Compact Disc test discs with opaque wedges, dots, and even simulated fingerprints to evaluate error protection, as well as other test tones and tone bursts for traditional measurements. More comprehensive specification and standardization is required to properly evaluate digital audio systems, and induce manufacturers to begin to optimize the performance of their products. Of course, specifications do not tell the whole story; advantages such as durability, convenience, and duplication without degradation clearly weigh in the favor of digital audio systems.

4.7 Alternative Digitization Methods

Although our discussions have focused on the classic digitization scheme, the linear PCM system, numerous other architectures are available. Depending on the specific application, methods such as floating point, block floating point, nonlinear and differential systems, and delta modulation offer good performance. Whereas a linear PCM system presents a set scale of equal quantization intervals into which the analog waveform is mapped, specialized systems offer modified or wholly new mapping techniques. One advantage of specialized techniques is often data compression, in which fewer bits are needed to encode the audio signal.

4.7-1 Floating Point Systems • A floating point system uses a PCM architecture modified to accept a scaling value; an example is shown in the block diagram in Fig. 4-30. Instead of a linear data word, the word is divided into two parts: the exponent (scalar) and mantissa (data word). Using the bits in the mantissa to represent the waveform's value, its proper amplitude is placed by the exponent bits; in other words, the exponent acts as a scalar which varies the gain of the signal entering the PCM A/D converter. By adjusting the gain of the signal the A/D is used more effectively; a low-level signal is boosted and a high-level signal is attenuated. For example, a 3 bit exponent might provide 8 different ranges for a 12 bit mantissa, as shown in Fig. 4-31. In this way 15 bits could be used to cover a large dynamic range, but only a 12 bit A/D converter would be required. One disadvantage is noise modulation; the quantization error varies with changes in the analog signal as the exponent shifts it up and down. Thus, the S/N ratio of the converter continually changes, as shown in Fig. 4-32. In this example, the 3 bit exponent switches in 6 dB increments, with 12 bit

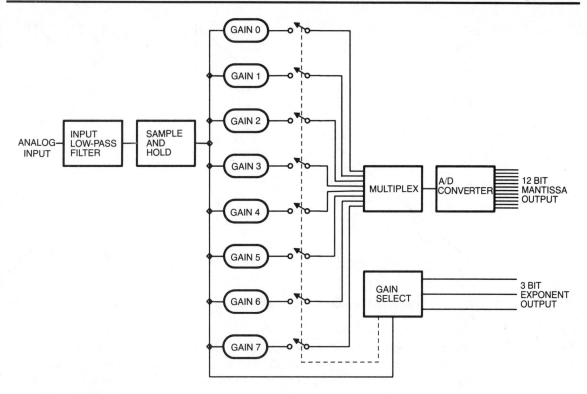

Fig. 4-30. **Floating point converter.**

mantissa. Another problem occurs with gain shifting; inaccuracies in calibration might present discontinuities as the different amplifiers are switched.

Another digitization architecture similar to the floating point system is the block floating point; its principle advantage is data compression. Thus, it is useful for transmission via satellite or other means. In block floating point, a full scale A/D precedes the scalar, a section of the analog waveform is converted to digital words, then a scale factor is calculated to represent the largest word in the block. The data is then scaled upward so the largest value is just below full scale, the number of bits needed to represent the signal is thus typically

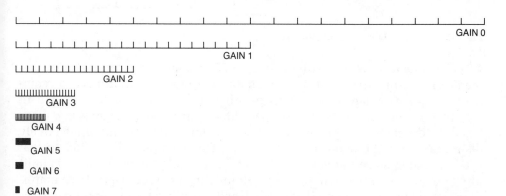

Fig. 4-31. **Quantization intervals vary with respect to the analog input signal in a floating point converter.**

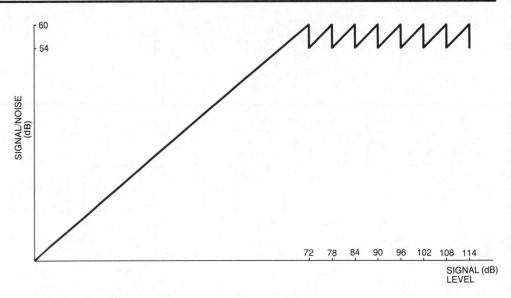

Fig. 4-32. **The signal to noise ratio varies in a floating point converter as the prescale gain varies.**

reduced. The data block is transmitted, along with the single scaling factor exponent. Upon reception, the block is properly rescaled. In the example in Fig. 4-33, 16 bit words are scaled to produce blocks of 10 bit words, each with one 3 bit exponent. Since only one exponent is required for the entire data block, data rate efficiency is high. System cost is also high, however, because of the full word A/D converter required.

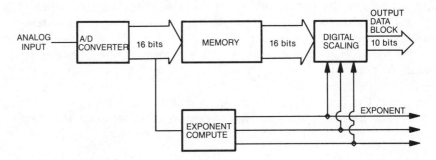

Fig. 4-33. **A block floating point encoder.**

4.7-2 Nonlinear Systems • With the linear PCM method, the quantization increments are evenly spaced across the amplitude range. In nonlinear, or compander systems, quantization steps of different sizes are used, as shown in Fig. 4-34. In general, the levels are spaced far apart for larger signals and close together for small signals, thus the quantization levels are more effectively distributed over the audio dynamic range. This technique is similar to that used in analog noise reduction; however, in this case the compander is built directly into the conversion logic thus eliminating tracking errors. High amplitude signals can be more easily encoded, and lower level signals will have less quantization noise; the result is a higher S/N ratio. Noise increases with large amplitude audio signals; however, the signal amplitude tends to mask this

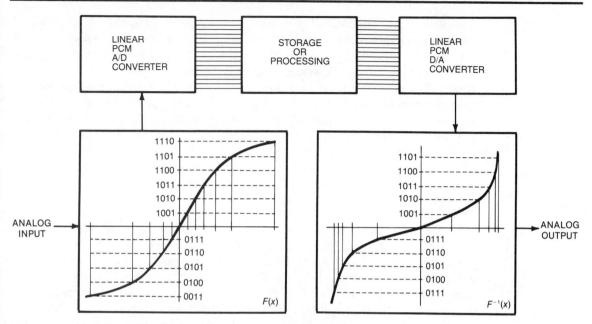

Fig. 4-34. **A nonlinear digitization system with companding elements.**

noise. Noise modulation audibility may be a problem for low frequency signals with quickly changing amplitudes.

4.7-3 Differential Systems • Differential PCM systems are significantly different from the linear PCM method. Intuitively we can see that we do not need to store the entire bulk of a waveform, only how it changes from instant to instant. In other words, rather than store long data words representing the entire amplitude of the signal, we need only store a few bits which represent the difference in amplitude between samples, as shown in Fig. 4-35. Positive or negative

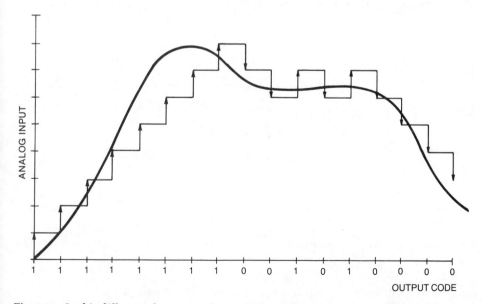

Fig. 4-35. **In this differential system, only one bit is used to encode the audio signal.**

transitions in the quantized waveform are used to encode the audio signal. A fast sampling rate is required to track the signal's transients. In the example in Fig. 4-36 the sampling rate is too slow to track the signal's rise time. An obvious advantage of a differential system is data compression. Differential encoding is a form of predictive encoding; a prediction for the upcoming sample is derived from the output data and the difference between the prediction and the input value is transmitted or stored. The decoder thus produces the prediction from the previous data, and with that plus the difference value the waveform is reconstructed sample by sample. This method reduces the number of bits required to encode an audio signal, but its success depends on the type of function used to derive the prediction signal and its ability to anticipate the changing signal.

4.7-4 Delta Modulation • Differential systems encode only the difference between the input signal and the prediction, as the sample rate increases the possible amount of change between samples decreases, and the encoding is more accurate. Delta modulation (DM) is a form of differential PCM which carries the method to the extreme. It employs a very high sampling rate so that only a one bit digitization of the difference signal is needed to encode the audio waveform, with little error. From both a hardware and operational standpoint, the method is simple and efficient. A block diagram of a delta modulation system is shown in Fig. 4-36. The prediction is compared to the present input and a one bit correcting word is generated at sample time; that is, the system determines if its error is positive or negative, and correspondingly moves its next value up or down one increment, always closer to the present value. At the output, the correction signal is decoded with an integrator. Only one correction can occur per sample interval, but a very fast rate could theoretically allow tracking of even a fast-transient audio waveform.

Delta modulation offers excellent error performance. In a linear PCM system, an uncorrected MSB would result in a large discontinuity in the signal. With delta modulation, there is no MSB, each bit merely tracks the difference

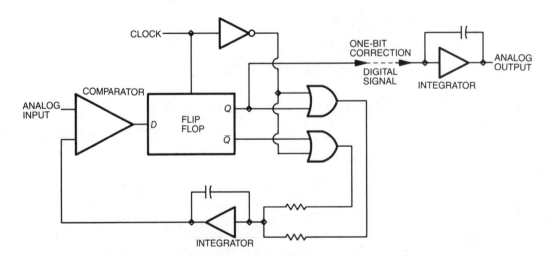

Fig. 4-36. **Block diagram of a delta modulation encoder and decoder.**

between samples thus inherently limiting the amount of error to that difference. The possibility of a tape degradation, however, necessitates error protection. Parity bits and interleaving are commonly employed.

On the practical side, delta modulation fails to perform well in high fidelity applications because sampling rate cannot be made fast enough. A single sign-changing bit cannot track a complex audio waveform; slew rate limitations yield transient distortion. From an informational standpoint, we can see that the method's design hampers its ability to encode audio information. A sampling rate of, for example, 100 MHz would permit encoding of frequencies up to 50 MHz, but most of that bandwidth is wasted because of the low frequency of audio signals. In other words, the informational encoding distribution of delta modulation is poor for audio applications. On the other hand, because of the large sampling rate, brick wall filters are not required. Very gentle filters with low phase shift may be used to adequately roll off audio frequencies well before half the sampling rate. Because of its limitations, delta modulation cannot be used for high fidelity applications, however several modified delta modulations offer more satisfactory results.

4.7-5 Adaptive Delta Modulation • To overcome transient response limitations of delta modulation, adaptive delta modulation (ADM) permits quantization increment size to be varied, to more quickly respond to the input signal. A block diagram of an adaptive delta modulation system encoder is shown in Fig. 4-37A. Algorithms examine the data then determine how to best adjust step size. For example, with a simple adaptive algorithm, a series of all-positive or all-negative difference bits would indicate a rapid change from the approximation, and the increment size would increase to follow suit, to more rapidly follow the change either positively or negatively. Alternating positive and negative difference bits would indicate good tracking and increment size would be reduced for even greater accuracy, as shown in Fig. 4-37B.

Hardware design is more complicated because the decoder must be synchronized to the varying strategy to properly recognize the changes in the step size. Also, it is difficult to change step size quickly and radically enough to accommodate sharp audio transients, and as high frequency and high amplitude signals demand large increments, quantization noise becomes larger too, thus producing noise modulation with the varying noise floor. In addition, it is difficult to inject a dither signal in an adaptive delta modulation scheme; since the increment sizes change, a fixed amount of dither is ineffective. Error feedback may be used to reduce in-band noise. A preemphasis filter characteristic may be used to reduce subjective noise in small amplitude signals, mask the change in noise with changing step size, and reduce low frequency noise in high amplitude, high frequency signals. As the audio slope increases, a control signal from the delta modulator, the same signal used to control step size, raises the frequency of the high-pass filter and attenuates low frequencies.

4.7-6 Companded Predictive Delta Modulation • Companded predictive delta modulation (CPDM) rejects adaptive delta modulation in favor of a compander delta modulation scheme. Instead of varying the step size in relation to the signal, the signal's amplitude may be varied prior to the constant step size delta modulator, to ensure against overloading the modulator. To reduce the quanti-

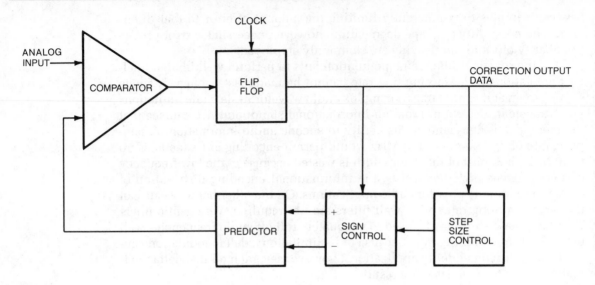

(A) Block diagram of adaptive delta modulation encoder.

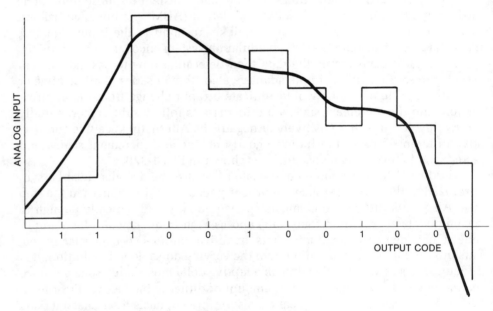

(B) Changes in step size are triggered by continuous 1s or 0s.

Fig. 4-37. Adaptive delta modulation encoder.

zation floor level, a linear predictive filter is used in which an algorithm utilizes many past samples to better predict the next sample. A block diagram of a companded predictive delta modulation system is shown in Fig. 4-38.

The companding subsystem consists of a digitally controlled amplifier in both the encoder and decoder sections for control of broadband signal gain. The bit stream itself controls both amplifiers to minimize tracking error. With the digitally controlled amplifiers, the signal is continually adjusted over a large range to best fit the fixed step size of the delta modulator. A transient "speed-up" circuit in the level sensing path allows faster gain reduction during

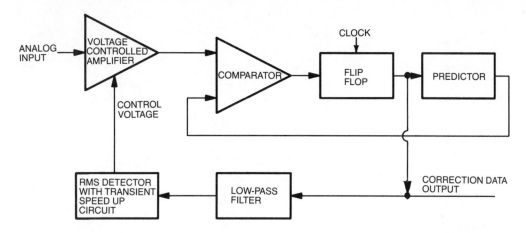

(A) CPDM encoder.

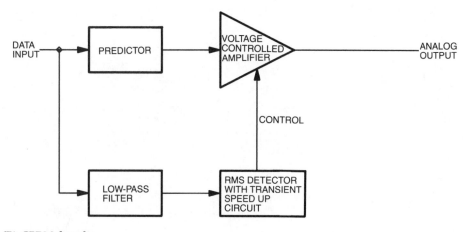

(B) CPDM decoder.

Fig. 4-38. Companded predictive delta modulation system.

audio transients. Strings of 1s or 0s indicate the onset of an overload and trigger compression of broadband gain to ensure that the transients are not clipped at the modulator. The speed of the gain change may be either fast or slow, depending on the musical dynamics. Spectral compression may be employed to reduce variations in spectral content; the circuit could reduce high frequencies when the input spectrum contains predominantly high frequencies, and boost high frequencies when the spectrum is weighted with low frequencies. The spectrum at the A/D converter is thus more nearly constant.

Companded predictive delta modulation, as well as other specialized designs offer alternatives to classic linear PCM design. They adhere to the same principles of sampling, quantization, error protection, and digital storage; however, their implementation is quite different. As digital audio technology evolves, other designs are certain to appear.

Chapter 5
Digital Audio Media

Introduction

Since the beginning of audio technology and the competition between cylinders and discs, many storage media have embodied audio software. Tape and optical disc (spelled with a "c" when referring to audio media) are the preeminent forms of storage for digital audio, as exemplified by stationary and rotary head tape recorders, and the Compact Disc. This chapter discusses in detail the two tape formats, and introduces other contenders for digital audio storage such as computer hard and floppy disk (spelled with a "k" when referring to computer media) and writable and erasable optical media. In addition, new forms of software dissemination, such as direct broadcast satellite and cable audio distribution systems, will be presented.

5.1 Digital Magnetic Recording

Magnetic recording has been a mainstay for analog storage of audio signals for over 40 years. Its ability to read, write, and erase has made it unique among storage media. With the advent and proliferation of digital computers, and their use of magnetic media for data storage, it is logical for digital audio data to be stored magnetically, using techniques pioneered by the computer industry. However, the large amounts of data contained in even a short musical selection place great demands on even the most sophisticated data storage and processing systems.

5.1-1 Recording Bandwidth • The bandwidth of a device measures the range of frequencies it is able to accommodate at an amplitude loss of typically no more than 3 dB, as shown in Fig. 5-1. In the case of an analog tape recorder, a bandwidth of 20 kHz (0 to 20 kHz) would be adequate since that band of audio frequencies is recorded directly on the tape, aided by a high frequency bias signal. However, a digital tape recorder requires a much larger bandwidth; the highest frequency to be recorded on the tape would be much higher than the 20 kHz audio frequency. Specifically, 20 kHz demands a sampling rate of 44.1 kHz. This sampling rate times the word length would yield 44.1 kHz X 16 or about 700 kilobits/second. If we add in overhead for data such as synchronization and error protection, the highest frequency to be recorded might be 1 MHz, or as high as 2 MHz. With modulation, the rate of transitions on the tape representing that frequency might be about 500 kHz. Thus, the recording bandwidth required of a digital tape recorder is ½ MHz (0 to 500 kHz), or about 25 times the bandwidth required of an analog recorder, given equal audio signal bandwidths.

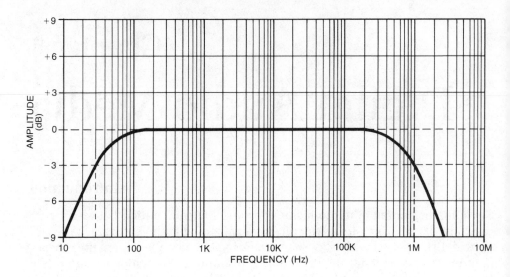

Fig. 5-1. **Recording bandwidth.**

While the recorded digital waveform may be highly distorted without affecting the audio signal quality, the mere problem of storing so many transitions is difficult. The problem of digital audio bandwidth can be understood when the data rate is considered; a stereo digital recording might require storage for a million bits for each second of sound. If a digital tape recorder has a lineal recording density of 20 kilobits per inch (kbpi), a large amount of magnetic tape is required for a lengthy audio recording. With professional digital multitrack recording of 24 or 32 audio channels, the problem is considerable. Thus, digital audio is successful at overcoming many of the limitations of analog tape recorders, but a large amount of digital data is required to store the same amount of audio information.

5.1-2 Digital Magnetic Tape • A magnetic tape is comprised of a plastic backing such as polyester coated with a thin layer of magnetic material such as gamma ferric oxide (Fe_2O_3). Gamma ferric oxide is a particle formation which is acicular (needle like) in shape, 10 to 25 microinches in length and 3 to 5 microinches in diameter. Each particle may be conceptually viewed as a magnet with a north and south pole, laid lengthwise along the tape, as shown in Fig. 5-2. To record information, an external magnetic field reverses the particles' polarity along the length of tape. In analog recording the relative net alignment of the particles represents the strength of the recorded signal thus a continuously variable change in analog amplitude may be stored. In a conventional digital recording, saturation recording is employed such that signals from the write (record) head cause entire regions of particles to be oriented either north pole first, or south pole first, depending on the digital 1 or 0 value of the data. Arbitrarily, a north-to-south pole change might be a digital 1, thus a south-to-north pole change would be a digital 0. During playback, the magnetic medium with its different pole-oriented regions passes before a read (reproduce) head which detects the changes in orientation, as shown in Fig. 5-3. The strength of the net magnetic changes recorded on the medium determines the medium's robustness; a strong recorded signal is desired because it can be read with less

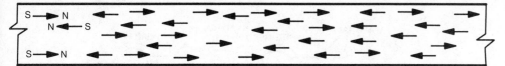

(A) Magnetic particles with a north and south pole are randomly oriented prior to recording, with no net magnetization.

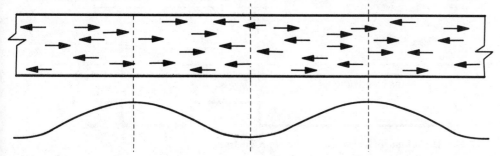

(B) Analog recording stores amplitude change through varying relative particle alignment.

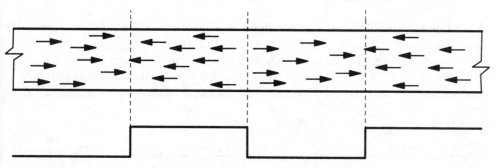

(C) In digital recording, flux reversals from N-S, S-N orientation represent digital data with saturation recording.

Fig. 5-2. **Magnetic tape recording.**

chance of error. Saturation recording ensures the greatest possible net variation in orientation of domains, hence it is robust.

Digital magnetic recording tape differs in several respects from that used in analog recording. Base thickness for analog tape is generally 1.15 mil to 1.45 mil (1 mil equals 1 thousandth of an inch); a thick base is required to minimize print-through of the recorded signal across tape layers while it is wound. With digital tape, higher track densities require precise tape-to-head contact, and since print-through is nonexistent, this is achieved with more flexible, and thus thinner tape, usually on the order of 0.80 to 0.88 mil. For analog tape, magnetic particle type is selected for noise, print-through, and distortion characteristics; coercivity, a measurement of magnetic force, is typically 300 to 400 oersteds. With digital tape, lineal density, the number of bits which may be recorded linearly on tape, as measured in kbpi, is the overriding concern. Lineal density in analog recording equivalent to 1.3 kbpi is required (15 ips and 20 kHz signal), however with digital storage (30 ips and 48 kHz sampling rate) the required density is 20 kbpi. To permit this density, particle types are chosen with higher magnetic energy levels which accommodate higher packing

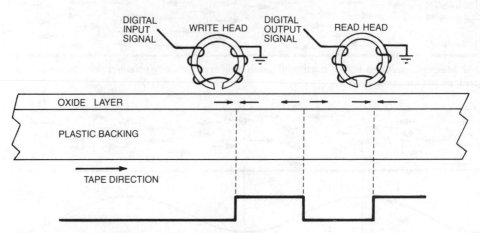

Fig. 5-3. **Digital magnetic recording and reproduction.**

densities. Thus, cobalt is often used to modify the gamma ferric oxide particles to achieve coercivity levels of 700 to 1000 Oersteds.

5.1-3 Longitudinal Magnetic Recording • Longitudinal recording is the method conventionally employed in which the transitions in the magnetically recorded waveform follow each other on the tape with N-S and S-N transitions recorded end to end, as shown in Fig. 5-4. High data density means short wavelengths,

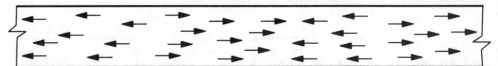

Fig. 5-4. **Longitudinal recording. Magnetic particles are placed end to end with N-S and S-N transitions.**

and as the particle poles come closer together, demagnetization occurs which leads to reduced output; there is thus a finite transition region which can be obtained for a given tape formulation. To obtain the high bandwidths required by digital recording higher tape speeds can be employed. Because of the resulting drawback of high tape consumption and short recording times, multiple tracks are often used. In other words, digital data from one audio channel may be written to multiple data tracks on the tape to achieve higher data density per area. The more tracks available for a signal, the greater the areal density, measured in $kbpi^2$, of the recording. However, the problem of tape defects is relatively larger with narrow tracks. As the areal density of the recording increases, error protection measures such as error correction codes and interleaving must become more sophisticated. It is the designer's challenge to maximize areal density; too low a density would dictate high tape consumption; too high a density would require additional error protection which would decrease the space available for data—a balance between the two must be achieved.

To increase the lineal and areal recording density of longitudinal recordings, thin-film heads may be employed. These heads use the same techniques used to manufacture hybrid integrated circuits, as shown in Fig. 5-5. Multiturn

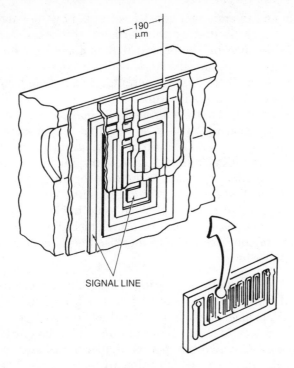

190
μm

SIGNAL LINE

Fig. 5-5. **Thin film heads use hybrid integrated circuit fabrication techniques.**

thin-film magnetic substrate heads are used for recording and magneto resistive (MR) thin-film heads are used for playback. Crosstalk between tracks is extremely low thus the tracks may be placed closer together. Signal-to-noise ratio is better than for conventional heads. The head's dimensions are precise because of the photolithography technique used to produce it.

Even with technical improvements and modifications, there is a limitation to the lineal density achievable by longitudinal magnetic recording techniques. The amount of information which can be stored on a track is limited by the number of magnetic particles which can be placed along the magnetic coating. A particle, like a magnet, should be cylindrically shaped with its length several times greater than its thickness. If this is not the case, self demagnetization reduces the strength of the particles as the poles neutralize each other, as shown in Fig. 5-6. Signal output is reduced as density is increased. Since the

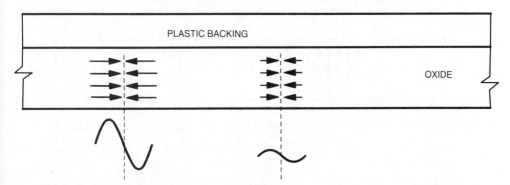

PLASTIC BACKING

OXIDE

Fig. 5-6. **Self demagnetization occurs at high lineal densities in longitudinal recording.**

particles are laid end to end, data density is limited by the length of the particles, which in turn is limited by the particles' thickness. For high density recording, the oxide layer on the tape or disk is made very thin to thus keep the particle thickness thin, and the track width is kept thin to similarly keep particle thickness thin. Together, this permits particle length to be decreased, thus allowing greater recording densities.

A limit is reached at a density of about 25,000 particles per inch; at this point the oxide layer coating must be so thin that the net magnetic output is quite low hence the signal-to-noise ratio of the medium, its ability to output an electrical signal above the background noise, is diminished. The only means to increase recording density per area is to increase the number of tracks recorded across an area. Sophisticated head positioning servo circuits and small head gaps in disk storage devices have achieved 1000 tracks per inch, and 10,000 tracks per inch appears possible. New recording techniques such as perpendicular and isotropic methods promise to increase the number of bits recordable per linear inch of medium.

5.1-4 Perpendicular Magnetic Recording • Perpendicular magnetic recording, sometimes called vertical or VR recording, differs from longitudinal recording in that the medium is magnetized at right angles to the surface, instead of along the surface, as shown in Fig. 5-7. To accomplish this, the particles are placed vertically in the magnetic medium, perpendicular to the surface. With longitudinal recording, the limiting factor of density is the length of the particles; however, with perpendicular recording, thinner particles yield greater densities—this is advantageous because greater density and thinner particles improves the length-to-thickness ratio and thus the magnetic strength of the medium. In other words, higher densities are apparently more robust. Whereas self demagnetization limits longitudinal recording to 25,000 bits per inch, research has suggested that perpendicular recording might be capable of a 500,000 bits per inch density, ultimately limited only by the size of the particles themselves. A digital audio system requiring storage for 800,000 bits per second would consume 40 inches per second of longitudinally recorded tape, but less than 2 inches per second with perpendicular recording techniques. Adjacent dipole fields defeat self-demagnetization thus permitting short wavelength recording.

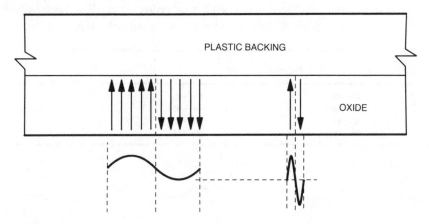

Fig. 5-7. **Perpendicular recording.**

To realize perpendicular recording, a suitable magnetic recording medium must be found; researchers have primarily used an alloy of chromium and cobalt in the form of hexagonal crystals. To place the layer on the medium, sputtering must be used, in which the medium is placed in a vacuum chamber and the chromium cobalt alloy is used as a cathode and struck with positive ions, and molecules of chromium cobalt are transferred to the medium. This is a costly and time-consuming process; more efficient methods will have to be developed for mass production of media. Even more elaborate perpendicular recording techniques use media with layers of both iron and nickel, and cobalt and chromium, and a double head design, as shown in Fig. 5-8. In this design, the secondary pole and Fe-Ni sublayer increases the longitudinal field gradient in the Co-Cr layer as recorded by the primary pole. Not all perpendicular recording systems employ a sublayer.

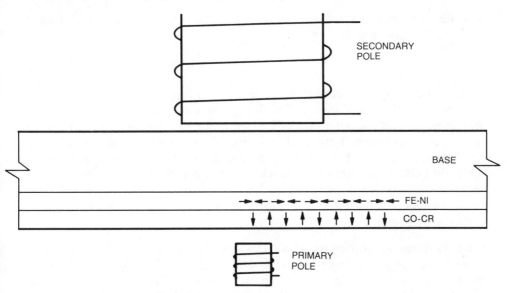

Fig. 5-8. **Head used for perpendicular recording.**

5.1-5 Isotropic Magnetic Recording • Isotropic recording takes advantage of the fact that an oxide layer can be magnetized in all directions; it records in longitudinal and perpendicular modes simultaneously. This is difficult to achieve because those two recorded fields tend to combine out of phase and thus produce an attenuated output signal. By matching head specifications to the properties of a tape formulation, it is possible to record those components in phase.

The isotropic head is designed so that the perpendicular record field erases longitudinal fields near the tape surface thus the tape is recorded internally with longitudinal fields, and with perpendicular fields at the tape surface, as shown in Fig. 5-9. Longitudinal components within the magnetic coating and perpendicular components near the surface comprise the recorded signal. Thick coatings help cover substrate defects. When the tape is played back, the fields are balanced so that the longitudinal field is dominant at low frequencies then reinforced by the perpendicular field at higher recorded frequencies. Transitions of 250,000 per inch have been achieved with isotropic recording techniques. The small head gaps employed in isotropic recording limit the

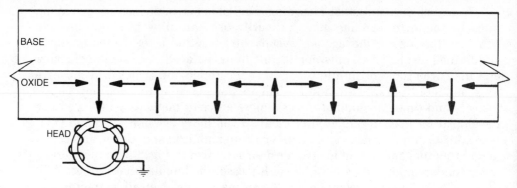

Fig. 5-9. Isotropic recording.

recording region to a small area at the following edge of the gap, thus there is essentially no peak shift.

5.2 Stationary Head Storage

Analog audio tape recording has primarily made use of stationary head recorders. Simplicity and low cost of design, long head life, facilitated editing, and compact and rugged construction are inherent advantages for analog stationary head recorders. With digital, the use of a stationary head design is more challenging because of the density of storage required. High tape speeds or multiple data tracks are necessitated. Rotary head digital audio recorder designs, discussed in the next section, using transports designed for storage of video signals, offer greater bandwidth, at the expense of more complicated editing procedures.

5.2-1 Features of Stationary Head Designs • Stationary head recorders offer several advantages over rotary head designs. Stationary head tapes are far easier to edit, perform punch-in and punch-out, and record and play back separate channels for synchronous recording; these are all important functions in professional multichannel recording applications. A rotary head tape recorder typically uses a cassette which must be edited electronically, and because all audio channels are multiplexed to one recorded track, it is not possible to work with individual tracks.

The recording engineer often uses synchronous recording techniques; some channels are recorded while other channels are being replayed in time synchronization. This overdubbing technique is used extensively in multitrack recordings. For synchronous recording, analog recorders use a record and reproduction head; old tracks are played back through the record head, while new tracks are simultaneously recorded on different record head tracks. If this was not the case, the performer would listen to the old tracks from the reproduction head, and record new tracks at the record head; because of the physical displacement between those heads the new tracks would be delayed in relation to the old tracks. With digital tape recorders it is difficult to intermix record and reproduction tracks in the same head. Thus, a digital tape recorder typically has two write (record) heads, one preceding the read (reproduce) head and used as a regular write head, and another following the read head, used in synchronization recording, as shown in Fig. 5-10. The write-1 and read heads

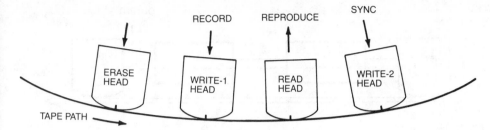

Fig. 5-10. **Head block for multichannel digital recorder.**

are used for read after write, and the read and write-2 heads are used for sync recording. Alternately two read heads and one write head could be used. The tracks played in synchronization are taken from the read head and delayed to coincide in time with the write head, as shown in Fig. 5-11. Signals from tracks 1, 2, and 3 (previously recorded) are taken from the read head and delayed to coincide with the writing of track 4.

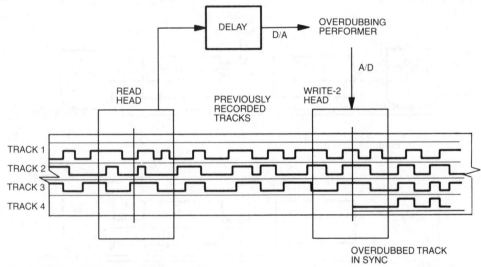

Fig. 5-11. **Synchronous overdub recording.**

Individual audio channels are often modified within the context of a recording by punching-in and punching-out, that is, as a channel is played back it is placed in the record mode at a certain point in the music to record new material, then taken out of the record mode; this procedure is illustrated in Fig. 5-12. Crossfading is used; this provides a smooth transition between the amplitudes of the two signals.

The professional recording engineer routinely edits analog tape with manual splicing techniques using a razor blade and splicing tape; with data interleaving and crossfade circuits to form a signal bridge across the area of disrupted data, razor blade splicing may be similarly accomplished on a stationary head digital recorder, as shown in Fig. 5-13. Much of this processing is permitted through the use of data buffers, as shown in Fig. 5-14. A delay between even and odd words, accomplished during encoding, permits tape splicing and crossfading at edit points; during playback the delay provides

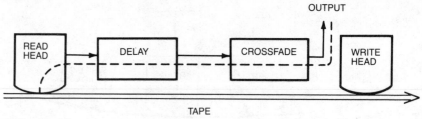

(A) Playback mode.

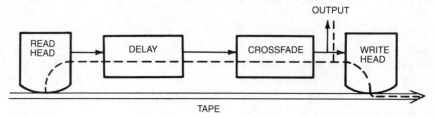

(B) Playback-record mode.

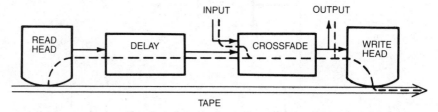

(C) Crossfade mode.

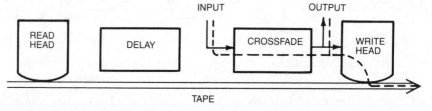

(D) Record mode.

Fig. 5-12. **Punch-in with a digital tape recorder. Punch-in procedure requires four steps.**

overlapping areas of information at the splice point. All of these functions are extremely difficult to accomplish with rotary head designs.

One disadvantage of stationary head designs is the added complexity of providing time synchronization in the bit stream (not to be confused with track-to-track synchronization). Video based rotary head systems use an inherently synchronized signal format whereas in stationary head designs a special synchronization word must be added to the bit stream periodically. This bit pattern is chosen to be uniquely different and recognizable from any audio data even when some of the synchronization word bits have been lost due to tape defects. This is a complicated procedure which adds to the cost of a stationary head system. The other features which a stationary head design supports, such as

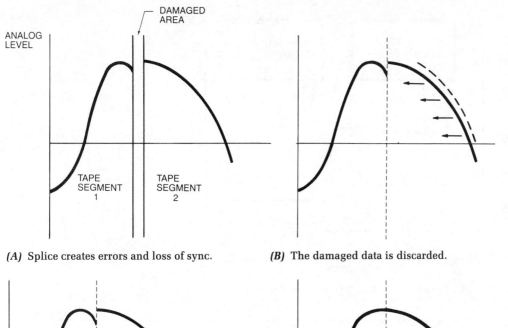

(A) Splice creates errors and loss of sync.

(B) The damaged data is discarded.

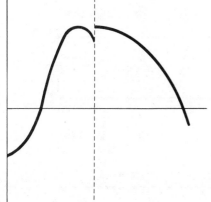

(C) The signals are joined.

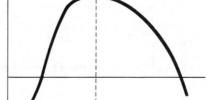

(D) Crossfading is used to create a smooth transition.

Fig. 5-13. **Digital tape splice processing.**

tape splicing and crossfading require additional circuits, as illustrated in the example of a stationary head reproduce section in Fig. 5-15.

5.2-2 DASH Format • The Digital Audio Stationary Head (DASH) format is an example of a strategy used for longitudinal digital audio recording. It is presently used for professional two-track and multitrack tape recorders. DASH is an attempt to establish a common format which will accommodate future technological improvements, such as thin-film heads, and still retain compatibility. Of course, reasonable tape consumption and robustness of recorded data are important design specifications.

The DASH format covers a wide range of applications from 2-track to 48-track recorders, using ¼ or ½ inch tape respectively, as shown in Table 5-1. The format has three versions depending on the tape speed specified: slow, medium, and fast; the number of tracks on the tape required to record one audio channel is four, two, and one, respectively, as shown in Fig. 5-16. For

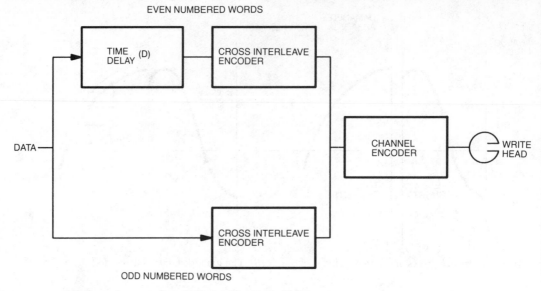

(A) Recorder recording section with even/odd delay.

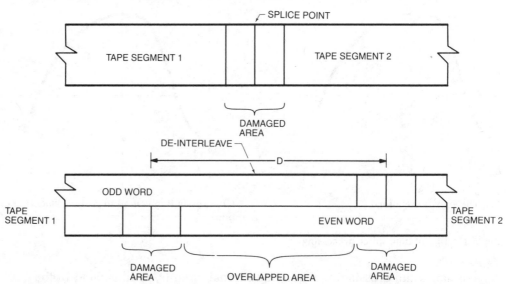

(B) Digital tape splicing with overlapping information.

Fig. 5-14. **Stationary head recorder tape splicing.**

example, to conserve tape in the slow speed version, the data is spread over four recorded tracks. The actual tape speeds for slow, medium, and fast formats vary according to the selection of a 48 or 44.1 kHz sampling rate, as shown in Table 5-2. A linear packing density is common to all versions; 38.4 kilobits/inch are recorded with 25.6 thousand flux reversals per inch, with a minimum wavelength of 78.2 mils and a maximum wavelength of 235 mils. The HDM-1 modulation code is used throughout.

Error protection provided for in the format is Cross Interleave Code (CIC), with interleave between even and odd input samples. Perfect correction is obtained for errors up to 8640 consecutive bits, good concealment for 33,982

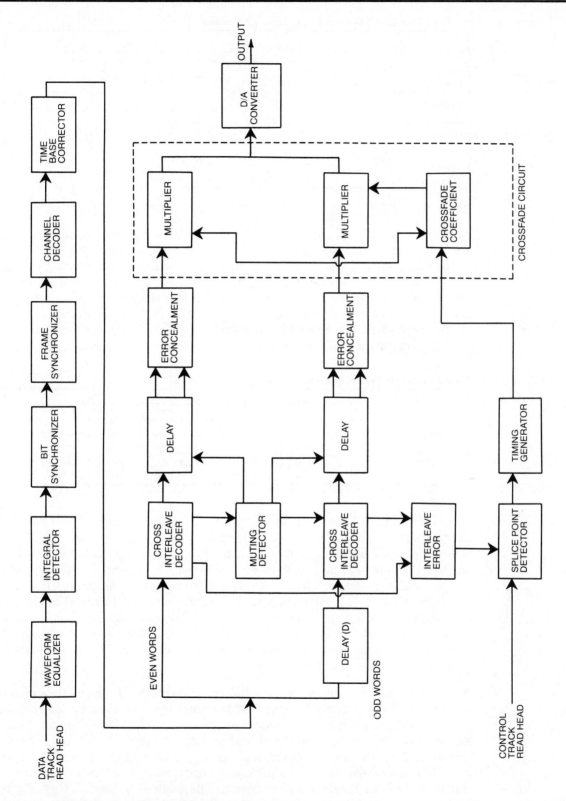

Fig. 5-15. Digital tape recorder reproduction section.

Table 5-1. DASH Track Density and Channel Number

Tape Width		1/4"		1/2"	
Track Density		Normal	Double	Normal	Double
Digital Tracks		8	16	24	48
Aux. Tracks		4	4	4	4
Digital Audio Channels	Fast	8	16	24	48
	Medium	—	8	—	24
	Slow	2	4	—	—

bits, and marginal concealment for 83,232 bits. Error protection is independent on each track in each version; even if one of the tracks is damaged, the error correction performance on the other tracks is not affected. Crossfading is provided for punch-in and punch-out, tape splicing, and electronic editing. A double density configuration of the format would utilize thin film heads for 48 tracks on a ½ inch tape. A stereo ¼ inch tape, 15 ips "Twin Recording" version of the DASH format is available for users who desire improved recorder performance in cueing and editing, afforded by the higher tape speed.

Table 5-2. DASH Tape Speed/Sampling Rate

Sampling Rate	Tape Speed		
	Fast	Medium	Slow
48 kHz	76.20 cm/s (30 ips)	38.10 cm/s (15 ips)	19.05 cm/s (7.5 ips)
44.1 kHz	70.01 cm/s (27.56 ips)	35.00 cm/s (13.78 ips)	17.50 cm/s (6.89 ips)

A stereo DASH recording on ¼ inch wide tape actually records 12 tracks on the tape, as shown in Fig. 5-17. Four auxiliary tracks are used for stereo analog cue track recording, a time code track, and a control track. Eight digital audio tracks are recorded; in the slow speed version, all eight would be used for two audio channels, while in the fast version, eight audio channels could be recorded. In a 24-track recording, 24 digital tracks are recorded thus only the fast speed version may be employed. The 4 auxiliary tracks are additionally recorded for a total of 28 tracks. In the double density version, 24 additional tracks would be sandwiched between the existing audio tracks to obtain a total of 48 tracks, with 4 auxiliary data tracks.

5.2-3 Summary of Applications • Stationary head digital recording thus offers many advantages, especially in terms of features important to recording professionals, such as editing and synchronous recording. High recording densities can be achieved through the use of multiple data tracks or higher tape speed; advances, such as thin film heads, will permit greater recording densities. The DASH format is an example of a stationary head format which incorporates the advantages of stationary head recording. The stationary head design is the choice for professional applications, in which flexibility in editing and over-

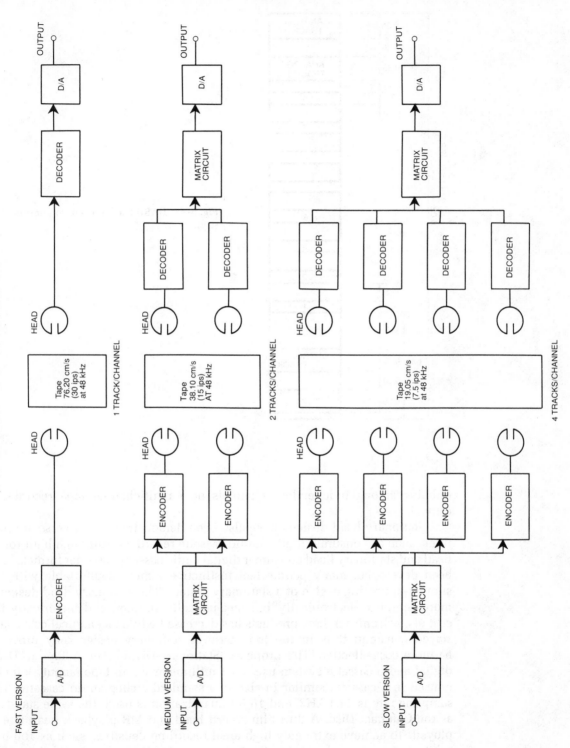

Fig. 5-16. **DASH format structure.**

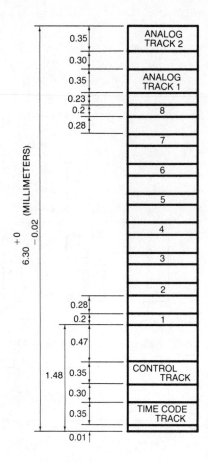

Fig. 5-17. **DASH track format (¼″, normal density).**

dubbing among independent channels of a multichannel tape recorder is critical.

Stationary head designs have also been demonstrated in diverse applications, such as consumer digital audio cassette recorders. Although both rotary head and stationary head consumer digital audio cassette recorder formats have been developed, many people feel instinctively more comfortable with the simpler-appearing design of a stationary system. The stationary head design is more complex electronically, but mechanically simpler, and historically the cost of electronics in new products has decreased while mechanical costs have stayed constant; thus in the long term, a stationary design may prove to be more cost effective. The proposed Stationary-Digital Audio Tape (S-DAT) digital audio cassette system uses 3.81 millimeter width tape (about ⅛ inch), housed in a cassette similar in size to a standard analog audio cassette. The sampling rate is 44.1 kHz, and 16 bit quantization is used, the same standard as the Compact Disc. A thin film record head and MR playback head is employed. To achieve extremely high areal recording densities, such as 100 megabits per square inch, 20 data tracks are recorded for each cassette side of the stereo signal. Because tape alignment is critical, a fixed-azimuth, edge-referenced guide system has been developed.

5.3 Rotary Head Storage

A full fidelity digital audio signal requires a 1 to 2 MHz storage bandwidth; since a video recorder has a bandwidth of over 4 MHz it is an ideal storage medium for a digital audio signal. Using a digital audio processor, an audio signal can be put into a video format and recorded onto a video tape recorder as if it were a video program. Upon playback, the signal is again converted to a digital audio bit stream and then to an analog waveform. Both consumer and professional type video recorders conforming to television standards (NTSC in the U.S. and Japan and PAL/SECAM in Europe) may be used to store the pseudo-video audio data.

5.3-1 Operation of a Video Tape Recorder • Although the information normally recorded on video tape represents the pixels, the points which comprise a video image, a processor may be used to encode audio signals into a video format thus permitting the use of video recorders in storing digital audio signals. To understand the limitations and opportunities of such a scheme, the operation of a video recorder must first be examined. As shown in Fig. 5-18,

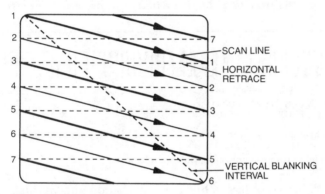

Fig. 5-18. **Television picture scanning.**

television picture is drawn on a cathode ray tube by a scanning process; 525 scan lines constitute the video image. Starting in the upper-left corner of the screen, the scanning spot moves across the screen diagonally, illuminating each of the pixels of red, blue, and green phosphorous with correct brightness and color information, then in horizontal flyback mode returns in a blanked fashion to begin another scanning line. To reduce flicker, interlacing is used to trace alternate scanning lines, that is, the odd field is illuminated, then it returns to the top of the screen in the vertical blanking interval to illuminate the even field thus two fields make one complete video picture. Depending on the standard being utilized, this process repeats 25 (PAL/SECAM) or 30 (NTSC) times per second, and as the information controlling the brightness and color changes, the display on the screen changes to effect a moving image. The time needed to scan one line (and return) is called a horizontal interval, and the time to scan from the top of the screen to the bottom (and return) is called a vertical interval. The actual video signal recorded on tape closely follows this format and is grouped in discontinuous parts, separated by horizontal and vertical retrace points, as shown in Fig. 5-19.

The high bandwidth required to record a video signal necessitates a high

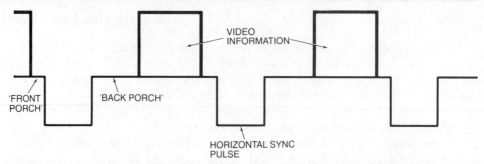

(A) Line synchronizing waveform showing horizontal sync pulses and television information.

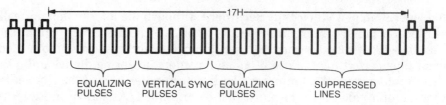

(B) Frame synchronizing waveform showing vertical blanking interval (even track).

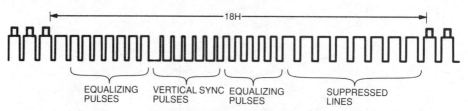

(C) Frame synchronizing waveform showing vertical blanking interval (odd track).

Fig. 5-19. Television frame synchronization waveforms.

tape speed. Video recorders achieve a high bandwidth by high head-to-tape speed. A rotary head is used with a technique called helical scanning. Two video heads are positioned on a head cylinder, since the cylinder rotates at 1800 revolutions per minute, a slow tape speed can be used. The tape is wrapped across the cylindrical drum, which revolves opposite to the tape direction, as shown in Fig. 5-20. Helical scan records diagonally aligned video tracks, alternating between even and odd fields. Since the tape is guided past the heads at an angle, each recorded track is placed diagonally across the tape width, as shown in Fig. 5-21. Each head lays down a track alternately on the tape. The discontinuities between tracks marks the vertical retrace points.

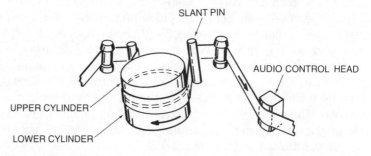

Fig. 5-20. Tape transport for helical scan in video cassette recorder.

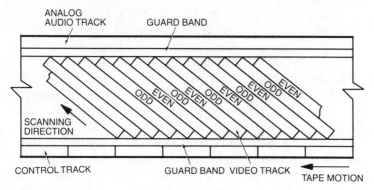

Fig. 5-21. **Recorded pattern of video cassette recorder.**

Thus, the video signal for one vertical period is recorded on each video track. There is a guard track between video tracks in the ¾ inch format; however, tracks are recorded side by side in ½ inch consumer video cassette recorders. This necessitates the use of a tracking control to adjust for proper alignment. Both the guard band or tracking control minimize effects from variations in the manufacturing specifications of tape which can cause misalignment between tape head and recorded tracks. A separate fixed head records a longitudinal track for analog audio accompanying video, as well as a control track which is used to phase lock the rotation of the head cylinder and the capstan servo, the electromechanical transport controller, to ensure accurate tape speed.

Assuming that the lowest frequency in a television signal is 25 Hz and the highest is 5 MHz, it is clear that a direct recording method is impossible due to bandwidth restrictions in the medium. This problem is solved in video recorders by modulating the signals onto carriers with frequency modulation. When a video signal modulates the carrier, the resultant sidebands determine the required recording bandwidth thus selection of an appropriate carrier frequency limits bandwidth requirements. With frequency modulation, a constant amplitude carrier is varied in frequency depending on the amplitude of the modulation signal, and the rate of variation similarly reflects the frequency of the modulating signal. For example, in the VHS format, the luminance (brightness) signal is FM modulated with a deviation from 3.4 to 4.4 MHz, while the chrominance (color) signal is modulated on a carrier centered at 629 kHz. In frequency modulated recorders, high frequencies are often boosted with a preemphasis circuit before recording, then equal but opposite postemphasis is used upon playback. This results in improved signal-to-noise ratio for high frequencies in the recorded signal. Frequency modulation, helical scan, and a sync pulse format thus form the foundations for video recording.

5.3-2 The Digital Audio Processor • To utilize a video recorder for digital audio storage, the digital audio signal must be processed to conform to the video signal format. Thus, the audio signal is transformed into a pseudo-video signal; to accomplish this, synchronization pulses appropriate for the television format are added to the digitally coded audio signals. In other words, the video recorder records what appears to it to be a television signal. The EIAJ digital audio processor format emulates NTSC or PAL/SECAM television signals so consumer digital audio processors can be connected to video cassette recorders, such as those using Beta or VHS format. Similarly, professional digital audio processors convert audio data to the NTSC or PAL/SECAM video formats, to be recorded on professional U-type video cassette recorders.

An audio processor functions in much the same way as the record and reproduce electronics of a stationary head digital tape recorder. The components of a PCM recorder, input low-pass filter, S/H circuit, A/D converter, error correction circuits, multiplexer, demultiplexer, D/A converter, and output filter, are all also present in an audio processor. Alternately, an audio processor could contain the circuitry required for delta modulation encoding and decoding. However, an audio processor must also contain circuitry to create the simulated video signal to permit storage on a video recorder. Several important operations take place. The pseudo-video signal must be given horizontal and vertical sync pulses, and head switching periods. This can be readily accomplished with the FM process used in video recording however the resulting discontinuity in the signal necessitates further processing. A general block diagram for a consumer PCM audio processor is shown in Fig. 5-22.

Video is recorded frame by frame in discrete lines whereas audio is comprised of continuous data. A video recorder divides the data into separate blocks. During the vertical sync pulse there is a gap equivalent in duration to 17 (odd tracks) or 18 (even tracks) scanning lines during which no data is recorded. Thus for audio recording, data time compression before recording, and data time expansion at playback, must be employed. The audio signal is directed through a record buffer memory so that during the discontinuities audio data may continuously flow into the processor. Similarly, on the output side, a reproduction buffer allows for continuous output flow of audio data even though the data arrives in the memory in separate blocks. In addition, because a video recorder only records one channel of data, all audio channels must be multiplexed into one channel. Buffers are used to store samples from alternate channels as the counterpart sample is being recorded on the video track.

The method of recording audio data on a rotary head system is different than for a stationary head system. A stationary head system typically records changes in the binary signal's polarity to designate a binary 1 or 0. However, in video recorders FM modulation is used so that changes in frequency represent 1s and 0s. The signals used to identify the end of a video horizontal or vertical line, called sync pulses, are represented with a third frequency. The recorded wavelengths are much shorter than the audio PCM wavelengths; the high bandwidth of the rotary head design supports this data overhead.

Because only one video tape recorder track is recorded at a time, a defect in the tape oxide, or an obstruction, such as a dust particle, could destroy a large number of successive samples. Thus, data is recorded with interleaving between samples. Once again, a buffer memory is used to permit this. Interleaving ensures that time adjacent samples will not be destroyed, instead the errors will be distributed in time, where they are easier to correct. Also for reasons of error protection, redundant samples are recorded. For example, the Cyclic Redundancy Check Code (CRCC) might be used; during recording the CRCC system records an error correcting word and a check word for every six words of audio data. During playback, code error is detected by monitoring check bits and the amplitude of the playback FM signal. When an error is discovered, error correction code is used to reconstruct the proper word. For more severe errors, linear interpolation is used to average the values of the

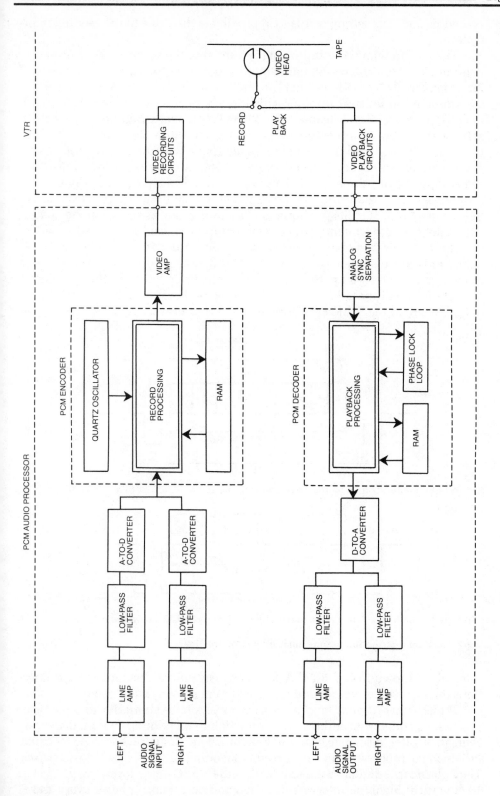

Fig. 5-22. **PCM digital audio processor system.**

preceding and succeeding words to approximate the value of the missing information.

The way in which the audio data is fitted into the video frame is primarily determined by the video scan rate. Since a video frame occupies $\frac{1}{30}$ of a second for a standard NTSC video signal (in the U.S. and Japan), the data for $\frac{1}{30}$ of a second of audio must fit into one video frame, that is, two video tracks. With the EIAJ format, a sampling rate of 44.05594 kHz (usually rounded off to 44.056 kHz) with 14 bit quantization permits a stereo audio signal with full error protection to be placed in one video frame. Using the video format, quantized audio data is placed on each horizontal scanning line instead of a television signal. Specifically, three 14 bit samples of audio data are recorded, along with a 16-bit CRCC word for drop-out detection and double parities for error correction, for a total of 168 bits per horizontal scanning line, as shown in Fig. 5-23A. Alternately, a 16 bit quantization signal could be placed in a video frame, at the cost of fewer bits given to error protection. In the 16 bit mode, six 16 bit data words are recorded, along with a 16 bit CRCC word; however, there is room for only one 16-bit parity word in the remaining interval, thus some error correction capability is lost. The 16 bit horizontal scanning line is illustrated in Fig. 5-23B. In the EIAJ/NTSC format, 262.5 H intervals comprise one vertical

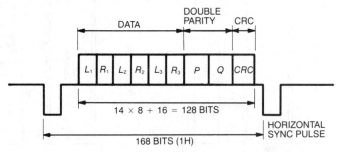

(A) In 14 bit mode, six data words, two parity words, and one CRC word are recorded.

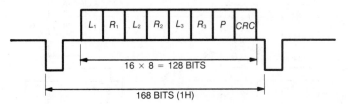

(B) In 16 bit mode, six data words, one parity word, and one CRC word are recorded.

Fig. 5-23. Data distribution for one horizontal scanning line.

interval, as shown in Fig. 5-24. A comprehensive block diagram and specifications for a consumer digital audio processor are shown in Fig. 5-25.

Professional digital audio processors operate similarly to consumer processors, converting audio signals into digital, then into pseudo video; however, they offer design features not included in their lower-priced cousins. Professional processors are used in the encoding process for Compact Discs. They place PCM digital audio data within the NTSC video format with 17 H or 18 H vertical blanking intervals (for even and odd tracks) placed every 245 H intervals, as shown in Fig. 5-26A. Each horizontal synchronizing pulse of

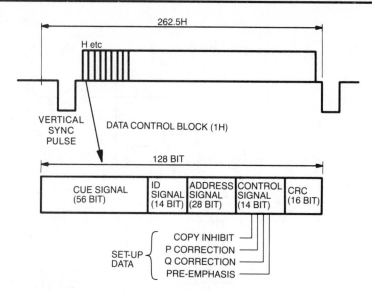

Fig. 5-24. **Data distribution for one vertical interval. Horizontal data blocks are separated by vertical sync pulses. The first H block is a control block.**

pseudo-video signal contains six 32-bit data words, as shown in Fig. 5-26B. The data contains almost 50 percent Cross Word Code redundancy for error protection. Together with interleaving the Cross Word Code can perfectly correct burst errors as large as 2240 bits (11.7 H) in one interleaved block of 6720 bits (35 H). Linear interpolation is used for errors as large as 4480 bits (23.3 H). Professional audio processors are often equipped with a manual preemphasis switch; it provides a 10 dB lift at 20 kHz, as shown in Fig. 5-26C. A code in the bit stream indicates preemphasis on/off so that deemphasis circuitry can be automatically activated when appropriate. Record circuits include preemphasis, anti-aliasing filter, sampler, A/D converter, error protection and interleaving, time compensation buffer, and video sync generation. Upon playback, the signal passes through a sync separator, time compensation buffer, D/A converter, anti-imaging filter, and deemphasis, as shown in the block diagram in Fig. 5-27. Specifications for a professional PCM digital audio processor are shown also.

5.3-3 Advantages and Disadvantages of Rotary Head Design • The use of a rotary head design for digital audio recorders offers the opportunity to achieve the high bandwidth required for digital audio storage. The narrow track width and high head-to-tape speed of rotary head designs yields high recording density and low tape consumption. The electronics required are relatively simple, since all the data is multiplexed to a single recorded track, redundancy of recording circuitry is greatly reduced. Data synchronization is simple to obtain in a rotary head design using a video format since the video sync pulses are an inherent part of the format. Because of mass production video recorders can be manufactured at low cost. Thus, a relatively sophisticated storage medium is widely available to both consumers and professionals.

Disadvantages of rotary head designs are related mainly to professional applications. Razor blade editing is not possible with a rotary head data track thus electronic editing must be employed using multiple tape recorders and

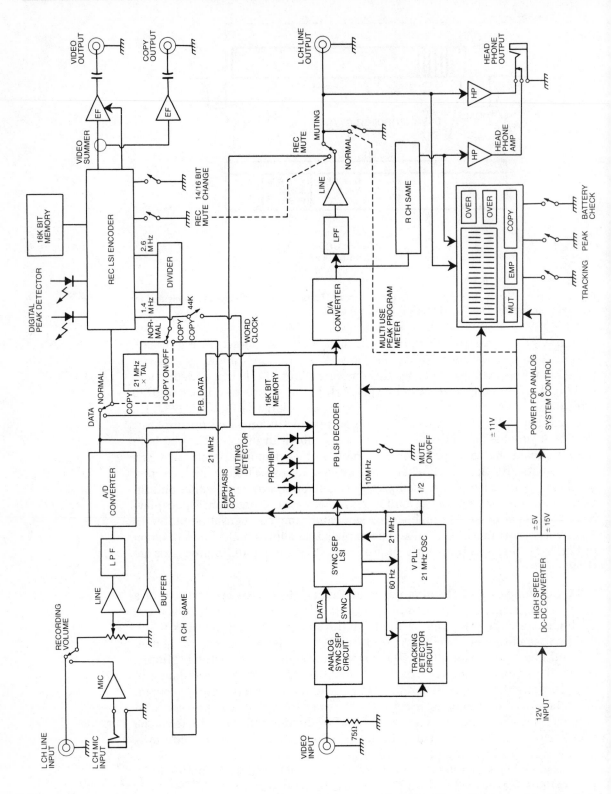

Fig. 5-25. Block diagram and specifications

associated controlling circuitry. Standard video editors are often not adequate since more precise subframe editing is often required in audio. Digital audio editors have been developed. Another disadvantage of rotary head design stems from the multiplexing of the channels, it is not possible to record and replay separate channels simultaneously. Similarly, punch-in and punch-out is not readily feasible.

Professional audio applications which require editing, multitrack and synchronous recording capabilities will probably retain the stationary head design for digital audio recorders. However, rotary head designs are highly successful for two track professional applications, especially as an encoding medium for the Compact Disc. Rotary head design is efficient for consumer formats, both as separate processor and recorder and self contained units. But using a video recorder for digital audio storage has an inherent drawback in that much extraneous information must be recorded because of the nature of the video format.

Specifications

System	PCM Encoding and Decoding with NTSC-compatible TV output
Number of Channels	Two
Sampling Rate	44.056 samples per second (each channel)
Quantization	14-bit linear PCM or 16-bit linear PCM
Coding Format	EIAJ-standard (14-bit) and 16-bit format
Bit Rate	128 bits per horizontal scan line (TVH) plus error-correction codes
Muting Condition	Dropout greater than 96 TVH
Equalization	50 μs pre-emphasis plus 15 μs de-emphasis
Frequency Response	10 Hz to 20,000 Hz ±0.5 dB (nominal)
	10 Hz to 20,000 Hz ±1 dB (guaranteed)
Dynamic Range	*14-bit format:*
(re 1 kHz, 0 dB, auto-	More than 88 dB (nominal)
mute on; 400-Hz to	More than 86 dB (guaranteed)
30-kHz bandpass filter)	*16-bit format:*
	More than 92 dB (nominal)
	More than 90 dB (guaranteed)
Total Harmonic Distortion	*14-bit format:*
	Less than 0.007% (nominal)
	16-bit format:
	Less than 0.005% (nominal)
	Less than 0.006% (guaranteed)
Channel Separation	Better than 80 dB
Wow and Flutter	Beneath measurable limits
Inputs (Line)	−10 dB, 40k ohms (for −15 dB recording level)
(Mic)	Low impedance
(Video)	1 volt P-P, 75 ohms
Outputs (Line)	−10 dB, load impedance greater than 10k ohms (from −15 dB recording level)
(Headphones)	−24 to −48 dB in 5 steps, low impedance
(Video)	1 volt P-P, 75 ohms
(Copy)	1 volt P-P, 75 ohms

for the (Sony PCM-F1) audio processor.

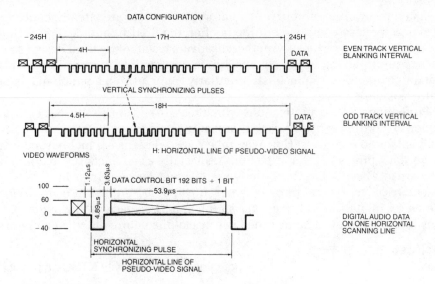

(A) Vertical and horizontal video waveforms.

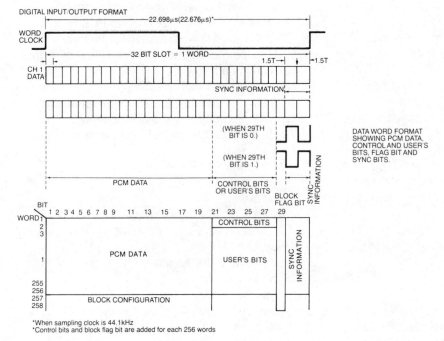

(B) Data format.

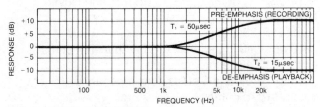

(C) Emphasis characteristics.

Fig. 5-26. Sony 1610 PCM digital audio processor format.

Horizontal and vertical sync pulses and FM modulation all contribute to data overhead. It is possible to use a rotary head transport optimized for digital audio which omits video sync pulses and FM and instead uses a direct recording format similar to that used in stationary head systems, yet enjoying the high data density of a rotary head design. Consumer digital audio cassette systems could thus efficiently employ a rotary head design. The rotary head design has already proven itself for consumer applications in the home video recorder market; from an engineering standpoint the rotary design lends itself to the high densities required for digital cassette recording. The Rotary-Digital Audio Tape (R-DAT) digital audio cassette format uses a transport similar to a video recorder; the helical scan technique permits slow tape speeds, and short tape lengths. The length of tape is one-fourth that of an analog cassette and the size of the shell is about half the size of an analog cassette yet it offers a playing time of over 2 hours. Tape width is 3.81 millimeters (about ⅛ inch); sampling rates are 32, 44.1, and 48 kHz, and quantization levels are 12 and 16 bit. A servo tape guidance system is required to maintain tape alignment relative to the heads. Rotary head cassette systems have achieved recording densities as high as 120 megabits per square inch. Because of the wide bandwidth, either digital audio or video images could be stored on the cassette.

5.4 Computer-Based Magnetic Storage

The technology used in digital audio is essentially an outgrowth of technology developed for the computer industry. Thus, it would seem likely that storage media designed for computer systems would also be applicable for audio storage. Although the storage specifications for the two applications differ, there is enough similarity to warrant compatible media. Thus, traditional computer storage media of disk pack and floppy disk may be utilized for audio storage.

5.4-1 Disk Pack Storage • Rather than develop special recorders for digital audio, it is possible to borrow equipment and techniques from mainframe computer technology. One audio recording system presently uses a minicomputer, standard digital instrumentation recorder, and custom converters, for two, four, and eight track digital recording. The tape recording is then transferred to one of two 300 megabyte removable disk pack drives, each capable of holding 21 minutes of stereo program, or correspondingly less time for multitrack programs. By loading the alternate drive while the other is in use, indefinitely long program times may be accommodated. The advantage of disk pack over tape is primarily evident in editing where the feature of random access is beneficial. Because of fast access times, a single disk pack can accomplish editing chores which would normally require two tape recorders. With a disk pack system, an edit list is sequentially followed and as one section of music is being played, the next section on the list can be located in memory and routed to an output buffer. In this way, a continuous data stream is output from disconnected sections of music on the disk. A block diagram of this system's architecture is shown in Fig. 5-28.

Special edit functions are available through programming; signal manipulations are limited only by appropriate software. In addition, since the editing is electronic, any edit may be independently previewed and modified before the final edits are selected. Even then, the original material remains unaffected;

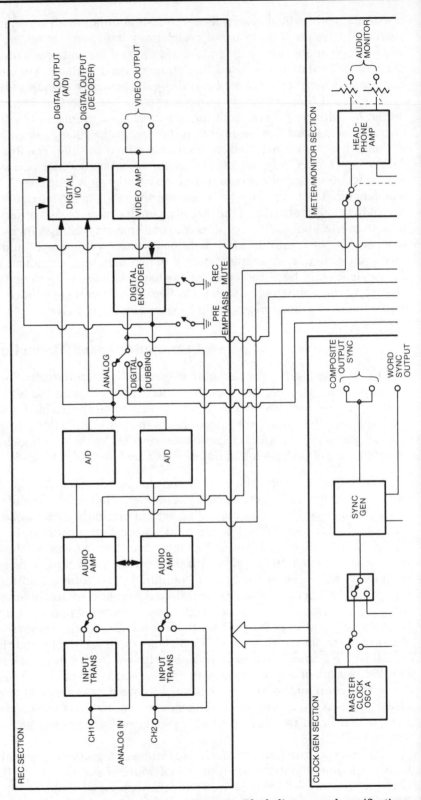

Fig. 5-27. **Block diagram and specifications**

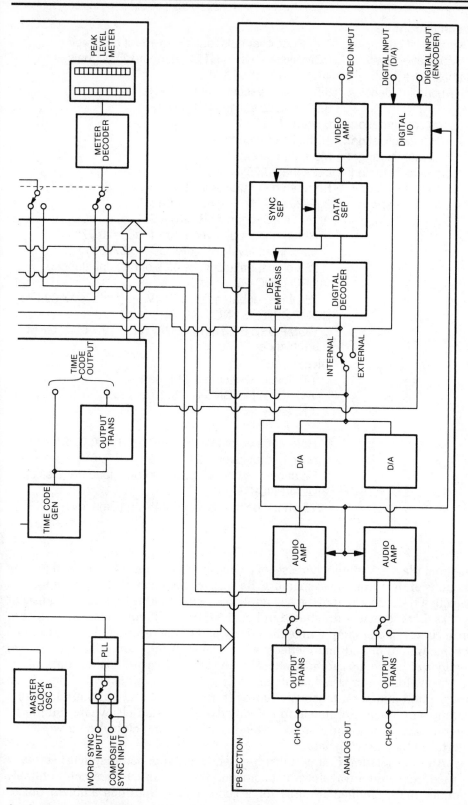

for the Sony 1610 audio processor.

(Continued on next page)

Specifications

Number of channels:	Two PCM channels, each with 16-bit accuracy
Modulation system:	PCM system using NTSC standard TV signals
Sampling frequencies:	44.056 kHz or 44.1 kHz
Recording density:	3.5795Mbits/second (44.056 kHz)
	3.5831Mbits/second (44.1 kHz)
Code configuration:	6 words in 1 TVH
Quantization:	16-bit linear quantization
Dynamic range:	More than 90dB
Harmonic distortion:	Less than 0.05%
Wow and flutter:	Beneath measurable limits
Frequency response:	20—20,000 Hz + 0.5, − 1.0dB
Inputs:	ANALOG(Cannon XLR-3-31). 2
	Reference input level +4dB (0dB = 0.775V)
	Max. input level + 24dB(0dB = 0.775V)
	25k ohms, balanced, or 4.7k ohms, unbalanced
	VIDEO (BNC-R) 1
	75 ohms, unbalanced, 0.714Vp-p (Data level 60 IRE)
	COMPOSITE SYNC(BNC-R). 1
	COMPOSITE SYNC (NEGATIVE), 4VP-P, 75 OHMS, unbalanced
	DIGITAL (BNC-R) 4
	TTL level, 32-slot serial format
	WORD SYNC(BNC-R). 1
	TTL level
Outputs:	ANALOG (Cannon XLR-3-3 2) 2
	Reference output level +4dB (0dB = 0.775V)
	Max. output level +24dB (0dB = 0.775V)
	Balanced, or unbalanced, 600-ohm load permissible
	VIDEO(BNC-R) 2
	75 ohms, unbalanced, 0.714Vp-p (Data level 60 IRE)

Fig. 5-27.

the final playback merely consists of a computer program selecting and joining sections of data from mass memory. Because of the block and gap data format, selective re-recording may be accomplished without affecting data in adjacent blocks. Data format is illustrated in Fig. 5-29. Any edit point may be located to an accuracy of one sample time thus every part of the recorded data is available for manipulation and many effects are possible. A butt edit concatenates two sections of music; by carefully choosing the edit points audible discontinuities can be avoided. A crossfade edit smoothly joins two sections by fading out the first section and simultaneously fading in the second section. Sound cloning permits any section of music to be copied and inserted elsewhere. Similarly, the duration of a sound may be extended by inserting cloned data words, or shortened by removing data.

An interpolation function can be used to provide better splices; software is used to examine waveform data surrounding the splice then compute new samples in the immediate area for a good match. Spurious ambient noises which have crept into the recording can be removed through editing, then an

TIME CODE 2
    Balanced (XLR-3-32), or unbalanced (BNC-R),
    600-ohm load permissible
    2.2V p-p, SMPTE time code
COMPOSITE SYNC (BNC-R) 2
    Composite sync (negative), 4 Vp-p, 75 ohms,
    unbalanced
DIGITAL (BNC-R) 4
    TTL level, 32-slot serial format
    WORD SYNC (BNC-R) 1
    TT level
HEADPHONES (Stereo phone) 1
    8 ohms

Power consumption:	Approx. 125W
Dimensions:	430 (W) x 280 (H) x 510 (D) mm (16.92 × 11.02 × 20.07″)
Weight:	Approx. 38 kg (83 lb 12 oz)
Supplied accessories:	Connecting cable with BNC connectors (2 pcs)
	Connecting cable with BNC and phono connectors (1 pc)
	Rack mounting adaptor (1 set)
	Extension board (1 pc)
	AC power cord (1 pc)
Usable VTRs:	BVU-100/110, BVU-200/200A/200B U-matic video cassette recorders
	BVH-500/500A, BVH-1000/1100/1100A 1″ video tape recorders
Recommended editing systems:	Regular system: BVU-200B (×2) + BVE-500A
	Advanced system: BVU-200B (×2) + DAE-1100

(Continued)

inaudibly compatible bridge may be built through interpolation. Amplitudes of entire sections or individual sounds may be varied. Edits are sometimes audible because of the jump in sound level; this can be overcome by digitally raising or lowering the amplitudes. Individual notes in a piece of music, perhaps played too softly or loudly, may be similarly adjusted to better fit the musical context. Time displacement may be altered as well; a premature entrance may be properly set back in time, or a late entrance moved forward. Ambience replacement permits removal or replacement of ambient information without altering the original musical material. All editing and data manipulations may be done with great precision because of a graphic visual display. The waveform itself is displayed and edit points selected by a video cursor with sample-to-sample accuracy. Display examples are shown in Fig. 5-30. A disk pack system permits highly flexible editing and postproduction processing. A master recording is input to the disk pack-based computer, processing is accomplished, then data is output to a second digital master tape as the instructions in the edit list are executed and appropriate musical data is taken from

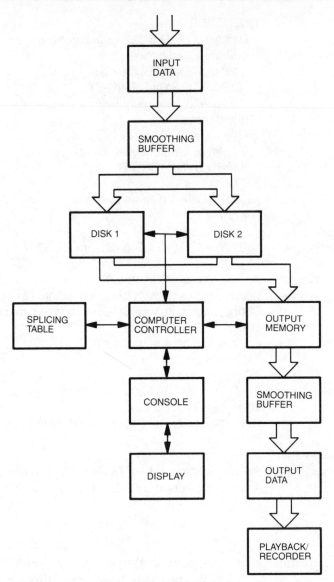

Fig. 5-28. **Soundstream digital random access editing system.**

the disk pack memory. The entire procedure may be accomplished by an operator with a computer terminal, video display and digital tablet for controlling cursor movement.

5.4-2 Floppy Disk Storage • Floppy disks are the primary means of permanent storage for microcomputers; however, conventional floppy disks are not practical storage media for digital audio data, indeed the magnitude of the storage requirements for digital audio can be illustrated by an attempt to use a conventional floppy disk. A disk storing one million bytes of information would hold about 10 seconds of unformatted stereo audio data. Even a hard disk storage system, such as used for micro and minicomputers, storing perhaps 25 Megabytes, would not be nearly sufficient. However, the random access ability, re-

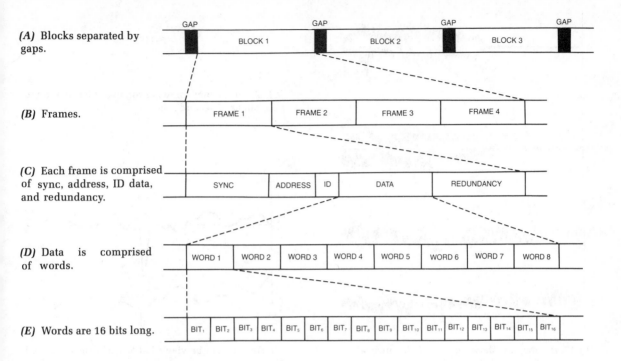

(A) Blocks separated by gaps.

(B) Frames.

(C) Each frame is comprised of sync, address, ID data, and redundancy.

(D) Data is comprised of words.

(E) Words are 16 bits long.

Fig. 5-29. **Block-coded digital format.**

cord ability, erase ability, and low cost of a disk have stimulated development of hard and floppy disk-based digital audio recorders.

One such floppy disk system has been proposed. It would use a 5¼ inch floppy disk to record and play 16 bit digital audio with consumer-priced equipment. The inexpensive vertical magnetization floppy disk would store a stereo program, achieving useful playing time with efficient modulation and packing of data. PCM data would be condensed by remaining zero levels, repeating instead of restating, and other data compression. The floppy disk drive would use a "gumball" type head, with a spindle speed of 360 revolutions per minute, and scan velocity of 2.6 feet per second. In addition, the unit would have the ability to interface with personal computers for music editing, synthesis, and restoration processing. Product development projects of this kind further illustrate the benefits of the application of computer technology to the problem of audio data storage.

5.5 Optical Storage

Magnetic media have been the preeminent means of digital storage for the last three decades. The computer revolution owes much of its success to magnetic media's ability to achieve high storage density at a low cost. The need for increased storage density has stimulated further advances in magnetic media; however, other researchers look to optical methods to achieve the high densities required for both future computer data and digital audio storage.

5.5-1 Overview of Optical Media • Magnetic media's advantage is its ability to be recorded, read, and erased with relative ease. With optical media, the tech-

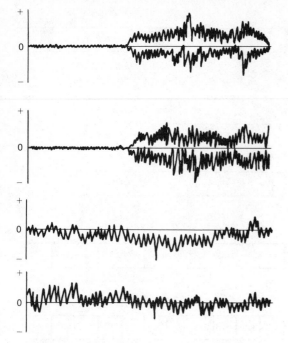

(A) Stereo display showing room ambience and musical attack, 2 seconds shown.

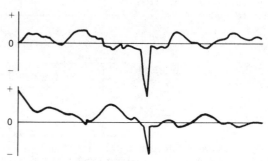

(B) Stereo display showing a click in musical program, ½₅ of a second shown.

(C) Display can be viewed at ½₅₀ of a second for editing or interpolation of click.

Fig. 5-30. **Example displays from Soundstream editor.**

nology required to erase and re-write data is quite sophisticated; thus first generation methods of optical storage will be read-only media, followed by write-once systems in which the user permanently records data, and finally by third generation fully erasable optical media. There are many varied types of magnetic media; similarly, many different types of optical storage systems will be designed depending upon the intended application. Optical disks of varying diameter, optical cassettes and reel-to-reel tapes and rectangular card media are all under varying stages of development.

Most optical storage systems operate with a mechanism in which a laser pickup shines on the media, and the reflected light is detected by a sensor and decoded to recover the carried data. To accomplish this, the media itself must present two states so that the change between them can be used to vary the reflected light, and thus the data be recognized by the sensor. For example, a reflective disk might have holes burned into its surface so that in one mode the laser light is almost entirely reflected back to the sensor, and in the other mode the light is dispersed, and the amount of reflected light is very small. There are many mechanisms available to achieve this, but questions of durability, density, and feasibility of mass production will ultimately point out the best methods.

5.5-2 Read-Only Optical Storage • The first generation of read-only storage is exemplified by the Compact Disc. A plastic disc is embossed with microscopic pits cut to a depth calculated to cancel and disperse the laser light of the pickup. The spiral track of pits holding modulated data can be followed by a tracking mechanism of a split laser beam. The reflective surface of the disc and the data pits are embedded between the transparent substrate and a thin protective layer. The effect of scratches and dust particles on the reading surface of the disc are minimized since they are separated from the data surface and thus made out of focus with respect to the laser beam focused on the inner reflective surface from underneath, as illustrated in Fig. 5-31. This type of media is particularly attractive because it can be economically mass produced. Furthermore, a disc can store over an hour of music with its 6 gigabit storage capacity. However, this is strictly a read-only media, and it could never be adapted to recording or erasing. The media is, however, open-ended with regard to the type of data to be encoded. Any digital data could be encoded thus not only music but computer software or video information, such as graphics and still pictures, could be disseminated via the Compact Disc in a system known as Compact Disc Read-Only Memory (CD-ROM) or Optical Read-Only Memory (OROM). Further discussion of the CD system is presented in Chapter 7.

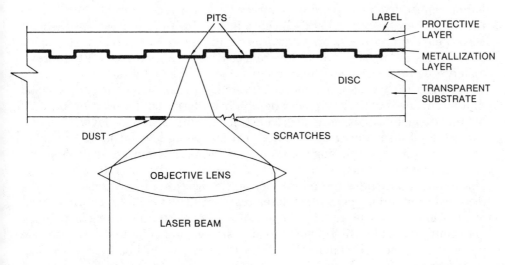

Fig. 5-31. **Compact Disc system data readout.**

5.5-3 Direct Read After Write Storage • Second generation optical storage systems will enable the user to record and play back his own data, without ability to erase and re-record. This permanent encoding is often referred to as Direct Read After Write (DRAW) storage. Several mechanisms are being researched, many systems use a laser writer to distort a thin metal film embedded in the media. Another mechanism called ablation uses a laser writer to burn holes in the thin film. Other mechanisms include lasers which cause bubbles or blisters to form in the media, or cause a phase change in the index of reflectivity of the media at the point where the laser strikes. Tellurium and tellurium alloys have been chosen by many manufacturers for the thin film because of their low melting points and high sensitivity.

All DRAW systems are write-once medias in which the data is nonerasable once it is recorded, in much the same way that a photograph is nonerasable. Although this limits its applications, DRAW can be an economical storage method. A disk might have a thousand times more storage capacity than a comparably sized magnetic disk; the cost per bit for optical storage is quite low.

An optical disk might store several gigabytes of data; a gigabyte (1024 megabytes) of capacity would record about an hour of audio data plus its error correction and formatting. One company is designing a system to store 500 optical DRAW disks, each with 4 gigabytes of memory. With such large amounts of storage capacity, the need to erase and record over unwanted data is alleviated; a user could simply keep recording new data, ignoring the unwanted. A high data rate of 5 to 20 million bits per second is expected.

One DRAW system currently uses an air-sandwich media, a 20 millimeter cavity is encapsulated by two plastic or glass substrates each 1.1 millimeter thick, then tellurium alloy recording layers each 30 nanometers thick are applied, as shown in Fig. 5-32. The media now being manufactured uses a 12 inch double-sided disk with a total capacity of 2.5 gigabytes using 0.7 micron diameter holes burned into the alloy film. Diode lasers are used to write data onto the disk, but data is read with a 2 milliwatt helium/neon gas laser.

Another DRAW system uses reflective and absorptive particles of silver halide embedded in a polymer matrix to form a recording medium called Drexon. When the silver halide layer is heated by a diode laser the particles absorb energy and the temperature rises until the upper polymer film melts and shrinks to create a hole of low reflectivity. A 12 inch disk with approximately 1.0 micron diameter holes holds 2.5 gigabytes on two sides.

A third DRAW system uses an irreversible phase change in a thin metal film to encode data; a low power helium/neon laser can be used to form 0.6 micron diameter spots on a disk constructed with a double layer structure consisting of a recording layer and a heat absorbing layer. The thin metallic film (recording layer) changes its physical property from an amorphous-to-crystalline phase when it is thermally heated by the laser to 170°C. The phase transition triples the reflectivity of the recording layer to delineate the recorded spots, thus permitting laser read-out of data. The recording layer, made of antimony selenium (Sb-Se) metallic film and the absorbing layer made of bismuth-tellurium (Bi-Te) metallic film, are evaporated onto a disk substrate made of poly-methyl-methacrylic (PMMA), as shown in Fig. 5-33. Specifications for prototype disks are shown also.

An alternative to thin metal film is the use of a polymer/dye binder bilayer medium; colored dyes in a plastic media over a reflective material are written with infrared (800 to 850 nanometer wavelength) light and read with

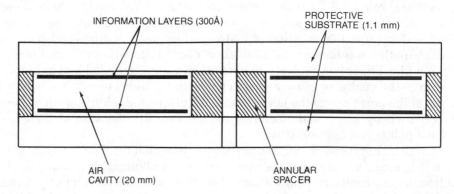

INFORMATION LAYERS (300Å)   PROTECTIVE SUBSTRATE (1.1 mm)

AIR CAVITY (20 mm)   ANNULAR SPACER

Fig. 5-32. **Cross section of air-sandwich Disc using thin-film tellurium recording surfaces. (Courtesy Philips)**

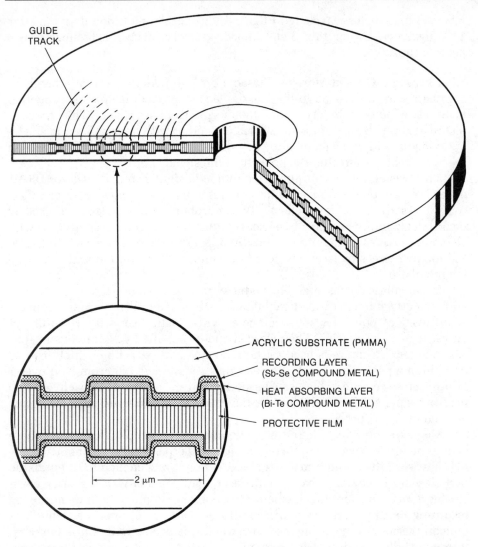

GUIDE
TRACK

ACRYLIC SUBSTRATE (PMMA)

RECORDING LAYER
(Sb-Se COMPOUND METAL)

HEAT ABSORBING LAYER
(Bi-Te COMPOUND METAL)

PROTECTIVE FILM

2 μm

	8-inch Disc	12-inch Disc
Disc diameter	200 mm	300 mm
Recording area		
Inside diameter	80 mm	100 mm
Outside diameter	180 mm	280 mm
Recording capacity (per side)	6×10^9 bits	15×10^9 bits
Track pitch	2 μm	2 μm
Number of tracks	25,000	45,000
Recording power of laser (1800 rpm)	Less than 6 mW	Less than 7 mW
Thickness of disc substrate	1.2 mm	1.2 mm
Substrate material	PMMA (Poly-Methyl Methacrylate)	
Substrate formation	Injection molding	

Fig. 5-33. **Direct Read After Write disc and specifications. (Courtesy Sony)**

red (633 nanometer wavelength) light. A 12 inch double-sided disc can store 11.2 gigabytes. Unlike thin metal film disks, polymer-dye technology uses a flexible substrate.

5.5-4 Erasable Optical Storage • Erasable optical media is currently being researched, and introduced to the marketplace. Two leading systems use either materials which exhibit crystalline-to-amorphous phase change when they are recorded at one temperature and erased at another with laser light, or magneto-optic media with a magnetic alloy recoding layer and a magnetic bias field in the laser light path to alter the polarity of the signal spots.

The crystalline-to-amorphous technique is similar to that used for DRAW disks; both are phase change media. For erasable media, typically a high reflectivity (crystalline) to low reflectivity (amorphous) phase change is used to record data, and the reverse to erase. A tellurium recoding layer, alloyed with elements, such as germanium and indium, may be directly recorded in DRAW fashion by increasing laser power to burn holes in the layer rather than changing the phase.

In one phase change erasable media, writing and reading are accomplished with an 830 nanometer wavelength laser, writing at 8 milliwatts and reading at 1 milliwatt of power. A 780 nanometer wavelength laser with 10 milliwatt power is used for erasing. The recorded spot is about 1 by 10 microns in size. A single lens structure is used for both erasing and recording, which may be done in one pass. The 8 inch diameter disk revolves at 1800 rpm, and over 1 million erase-rewrite cycles can be performed on a disk. A capacity of 700 megabytes per side has been achieved. Both erasable and write-once disks may be used in the recorder.

Magneto-optic techniques are slightly more complicated than phase change methods. Magneto-optic recording uses perpendicular magnetic recording; however, the applied magnetic field is not strong enough to orient oxide particles until they are heated with laser light. In this way the laser beam creates a recorded spot much smaller than otherwise possible thus increasing recording density. Terbium and iron alloyed with gadolinium or bismuth is typically used as a recording medium; when it is heated to its Curie temperature its coercivity approaches zero and its oxide particles are easily reverse-oriented by the magnetic field before returning to normal temperature. A coil is wrapped about the laser lens structure to produce the magnetic field. Its perpendicular alignment is assisted by a metal plate located on the opposite side of the disk. The Kerr (or Faraday) effect is used to read data; the plane of polarized light is rotated when it strikes a magnetized material. The rotation from the reverse oriented particles in contrast to the normal-oriented particles produces a modulated light beam which carries the data, as shown in Fig. 5-34. Erasure uses the same method as recording, but with a reversed magnetic field.

One erasable magneto-optical disk system using a 120 millimeter diameter disk (the same size as a Compact Disc), has been used to digitally record an hour of music which can be erased and re-recorded. A photopolymer layer with a pre-grooved surface is deposited on a glass substrate; a dielectric layer separates it from the amorphous recording layer of terbium, gadolinium, and iron (Tb,Gd, and Fe). A top protective layer completes the disk construction, and the disk is accessed through the glass substrate, as shown in Fig. 5-35.

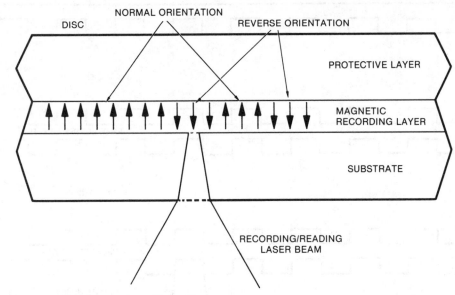

Fig. 5-34. **Magneto-optic erasable medium.**

Recording is accomplished by spot heating of the recording layer to decrease the coercivity of the alloy. Simultaneously, a magnetic field perpendicularly magnetizes the layer; this rotates the plane of polarization of plane-polarized light used to read the disk. The recorder/reader is shown in Fig. 5-36. The light source is an AlGaAs laser with 850 nanometer wavelength; the beam is pulsed during recording for a duration of 50 nanoseconds at 250 nanosecond intervals. With the help of a magnetic coil underneath the disk, oxide particles within the groove are oriented by the light pulse. To read the disk, the pulsed laser light reflects from the disk and the difference in rotation of the plane of polarization of the reflected light is read as intensity modulation. Erasure is accomplished by uniformly magnetizing the surface, then re-recording. With minor modifications, such an erasable recorder/reader could be used to read standard Compact Discs as well.

Another erasable magneto-optic disk system suitable for digital audio applications uses a recording layer of a thin amorphous magnetic film of terbium, iron, and cobalt (Tb-Fe-Co) recorded with perpendicular magnetization. A protective layer and outer poly-methyl methacrylate (PMMA) layer forms the final surface of the disk. A semiconductor laser beam is used for recording, reading and erasure. A carrier-to-noise ratio of over 50 dB at 1 megahertz has been reported. Disk construction and specifications for various prototype disks are shown in Fig. 5-37.

Other types of optical storage, still exotic, are copper sulfate in glass technology in which formations of bumps at a density of 10^8 per square centimeter have been achieved, organic dye techniques, cryogenic frequency domain storage, surface texturing techniques, and spectral-hole burning in crystals. Because of the complexity of the technology, erasable optical media will initially be expensive. However, optical storage technology, in all of its three realizations of read-only, write-once, and erasable, will be increasingly abundant and cost effective in the future.

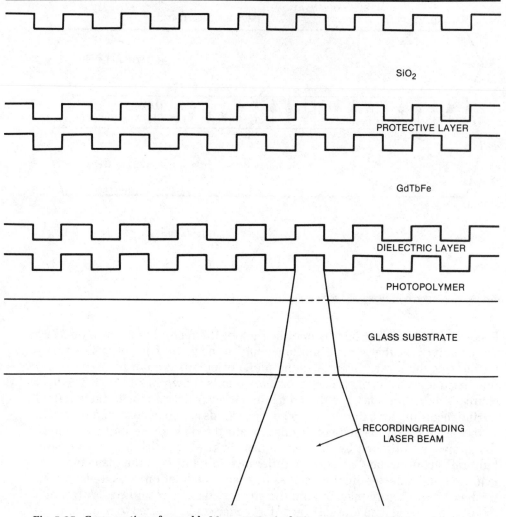

SiO$_2$

PROTECTIVE LAYER

GdTbFe

DIELECTRIC LAYER

PHOTOPOLYMER

GLASS SUBSTRATE

RECORDING/READING
LASER BEAM

Fig. 5-35. **Cross section of erasable Magneto-Optical Disc. (Courtesy Philips)**

5.5-5 Laser Card Storage • One limitation of all existing audio storage media, analog or digital, is that of moving media. The mechanical task of rotating disks and positioning a pickup, or moving tape past stationary or rotary heads places an inherent limit on cost and size. While it will be some time before solid-state memory will have sufficient capacity with small size and cost to accomplish audio storage, other projects are underway. A laser card storage medium has been proposed. A thin plastic card would hold recorded material and offer low cost digital playback for the consumer. A laser beam would be deflected across the card face to recover data from the card. A write-once feature would permit data to be written to the card and read back with a laser; however, it could not be erased. Because the laser beam would be deflected electrically, the card's position would be fixed, thus cost and size could be quite low. Manufacturers are now at work on product development and such systems could represent the next generation of digital audio storage mediums.

A write-once optical card is being offered, using the silver halide particle

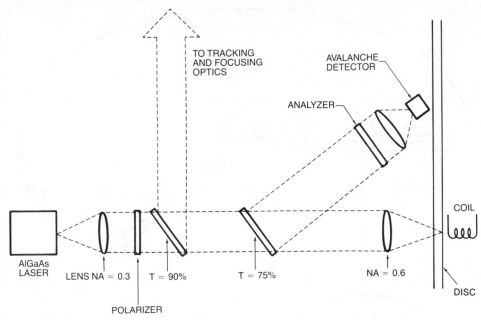

Fig. 5-36. Optics for Magneto-Optical Disc recorder/reader. (Courtesy Philips)

technique. With a credit card-sized card with both sides coated with Drexon, storage capacity could approach 100 megabytes. A semiconductor laser can be used to write data and a diode laser or simple photodetector array can be used to read data. It is conceivable that the capacity of such a system could be increased to point where the large amounts of data required for long-duration audio programs could be accommodated.

5.6 Transmission Media

Rather than distribute music and other data through individual manufactured media (e.g., records and tapes), a more efficient approach might be the use of transmission media. With letters and recordings, it isn't the piece of paper or plastic disc we are interested in, it is the information contained in those media which concern us. In the future, it is likely that much of our information, whether it be electronic mail or recorded music, will come to us as transmitted data, to be used in real time, or placed in private storage. Direct broadcast satellites, cable and interactive systems are examples of new transmission media.

5.6-1 Direct Broadcast Satellite • Direct Broadcast Satellite (DBS) is a system in which households equipped with a small parabolic antenna and tuner receive broadcasts directly from a satellite parked in a geostationary orbit. The satellite receives transmissions from ground stations and relays them so that individuals not otherwise in a reception area can receive high quality transmissions of television and digital audio signals. The receiving system is comprised of an offset parabolic antenna with a corrugated ring, electric wave collector designed to catch the microwave signals sent by the satellite, and a converter mounted at the antenna's focal point to convert the microwave signal to a lower frequency FM signal. Because of high sensitivity of these devices,

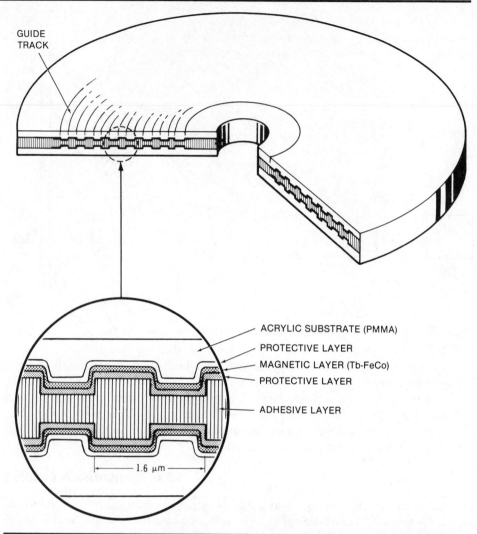

	5-inch disc	8-inch disc	12-inch disc
No. of tracks (one side)	16,000	28,000	45,000
Track pitch	1.6 μm	1.6 μm	1.6 μm
Recording capacity	4×10^9 bits	12×10^9 bits	30×10^9 bits
Recording & erasure power of laser at 1800 rpm	Less than 3 mW	Less than 5 mW	Less than 7 mW
Reading power	1 mW	Less than 2 mW	Less than 2 mW
Thickness of disc substrate	1.2 mm	1.2 mm	1.2 mm
Disc substrate material	PMMA	PMMA	PMMA

Fig. 5-37. **Magneto-Optical Disc and specifications. (Courtesy Sony)**

the parabolic antenna may be as small as ½ meter in diameter. The units are mounted outside the home on porch or roof, and are manually aligned with a diagnostic display showing received signal strength. Inside the home, the third component of the receiving system, a phase lock loop tuner, demodulates the FM signal from the converter into video and audio signals suitable for a home television or stereo; using circuitry similar to that found in Compact Disc players, the tuner may decode PCM audio data for high fidelity music repro-duction through home stereo systems. A system block diagram is shown in Fig. 5-38. For areas not centrally located in the satellite's "footprint" broadcast area, larger antennas of a meter or more in diameter may be used for favorable reception.

The wavelength used for direct broadcast satellite systems is known as Super High Frequency (SHF) and is higher in frequency than the VHF and UHF channels used for conventional television broadcasting. The broadcast spec-trum is shown in Fig. 5-39. In addition, the broadcast specifications for SHF

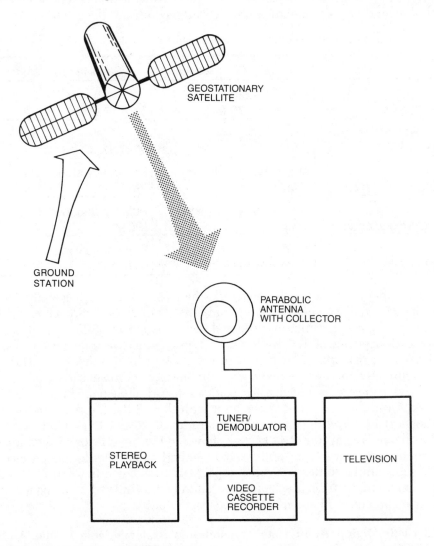

Fig. 5-38. **Direct broadcast satellite system.**

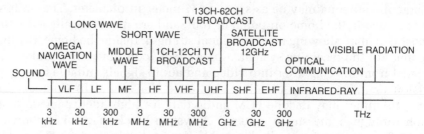

Fig. 5-39. **Frequency spectrum showing SHF satellite broadcast band.**

differ from those of conventional television, as shown in Chart 5-1. The 12 gigahertz signal's wavelength is 2.5 centimeters long and its frequency band is often called the quasi-millimeter wave. The signal has a line-of-sight character-istic similar to that of visible light, thus it is highly directional; when originat-ing from a high altitude, it can radiate over a wide area and can be received with a small diameter antenna if properly aligned toward the satellite.

Chart 5-1. **Comparison Between Television and Satellite Broadcast**

	current TV broadcast	Satellite broadcast
Frequency	VHF 90–222 MHz UHF 470–770 MHz	SHF 12 GHz
Frequency Bandwidth	6 MHz	27 MHz
Video modulation	Amplitude Modulation (AM)	Frequency Modulation (FM)
Audio modulation	Frequency Modulation (FM)	Pulse coded modulation (PCM)
Polarization	H. linear Polarization or V. linear Polarization	Right-hand circular polarization

The Japanese broadcasting network, the NHK, currently has a DBS satel-lite, the BS-IIa, at 110 degrees East at an altitude of 36,000 kilometers, broad-casting a 12 gigahertz microwave signal over Japan. The BS-IIa satellite has a transmission power of 100 watts for its two channels, a transmission band-width of 27 megahertz, video bandwidth of 4.5 megahertz, video S/N ratio with a 75 centimeter antenna in Tokyo is 43 dB, and an audio subcarrier is located at 5.73 megahertz. Its broadcast specifications are shown in Fig. 5-40. Two modes of PCM audio may be broadcast with the BS-IIa; the A mode offers 4 channels of 10 bits with compression, sampling at 32 kHz, the B mode has two 16 bit channels sampling at 48 kHz, as shown in Fig. 5-41. Error correction and control data accompany the audio signals in both modes. Performance specifi-cations are similar to those of the Compact Disc, as shown in Chart 5-2. In the future, more than 10 channels of audio could be transmitted by using a single television broadcast channel of the broadcast satellite.

5.6-2 Cable Digital Audio/Data Transmission System • Cable Digital Audio/ Data Transmission System (CADA) is a broadcast system for digital audio and

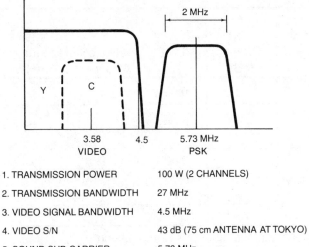

1. TRANSMISSION POWER   100 W (2 CHANNELS)

2. TRANSMISSION BANDWIDTH   27 MHz

3. VIDEO SIGNAL BANDWIDTH   4.5 MHz

4. VIDEO S/N   43 dB (75 cm ANTENNA AT TOKYO)

5. SOUND SUB-CARRIER   5.73 MHz

6. NUMBER OF SOUND CHANNELS   A MODE : 4 CHANNELS
                                      B MODE : 2 CHANNELS

Fig. 5-40. **Specifications for NHK BS-IIa DBS satellite.**

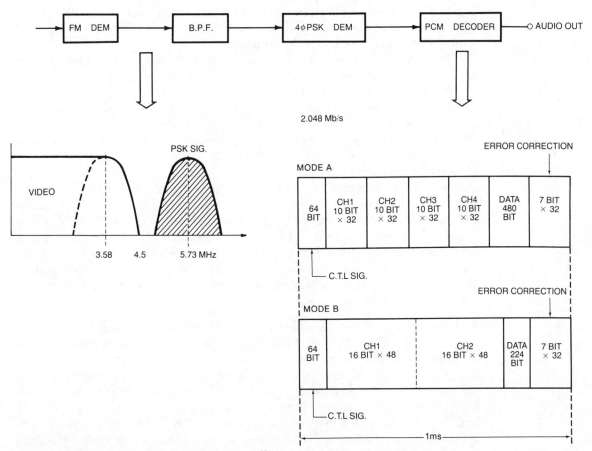

Fig. 5-41. **PCM audio specifications of BS-IIa satellite.**

Chart 5-2. Comparison of Audio Parameters of BS-IIa Satellite with Conventional Media

		Audio Frequency Range (kHz)	Dynamic Range (dB)	T.H.D (%)	No. of Channels	Sampling Frequency (KHz)	Quantization (bit)
BS IIa (D B S JAPAN)	Mode A	~15	84	0.08	4	32	10*
	Mode B	~20	95	0.003	2	48	16
Compact Disc		~20	95	0.003	2	44.1	16
Conventional TV Broadcast (EIAJ)		~15 main (14) sub	60	0.7 (1.5)	2	—	—
FM Radio		~15	75	0.1	2	—	—

*10/14 bit Quasi Instantaneous Compression

other data using existing cable television (CATV) lines of coaxial cable and optical fibers. The digital audio and data is transmitted using a frequency range equivalent to one arbitrarily assigned CATV channel. Such a system has features and specifications superior to a normal FM broadcast system in terms of quality, flexibility, and performance. By using error correction, the CADA system can maintain high quality even under adverse cable conditions when standard CATV signals can no longer transmit.

The CADA system would be compatible with any existing cable system with one open channel. High quality audio signals, such as those originating in digital form from Compact Discs, as well as facsimile data, still video pictures, and computer software could be programmed. One proposed system has four independent data channels within the single television channel; each data channel may select a mode from one of four modes, as shown in Fig. 5-42. The 4 modes could provide a stereo 44.1 kHz, 16 bit signal, 2 stereo 44.1 kHz 8 bit channels, 8 monaural 22.05 kilohertz 8 bit channels, or 1 stereo 44.1 kHz 8 bit channel and 4 monaural 22.05 kHz 8 bit channels. All channel modes would contain 7 bits of error correction code; four service bits and eight synchronization bits are also attached, forming a word consisting of 168 bits. With data channel formats, such as the one proposed previously, a very flexible system design is possible, with actual implementation depending on the demand. For example, using the four modes, a household could receive one stereo Compact Disc format channel, 32 channels of monaural low fidelity audio, facsimile data, and computer software.

5.6-3 Interactive Media • A wide variety of distribution media are sure to arise to service the increasing flow of information. However, rather than receive only the information chosen for mass distribution (in a world where the amount of information is much larger than a directed system could ever disseminate), individuals will be able to retrieve data selectively through interactive systems. Using cable, television, telephone and computer networks, users will be able to communicate with a central data source for a wide variety of information and services. In addition, it is realistic to expect that music recordings, and live concerts, will be available from the central data source.

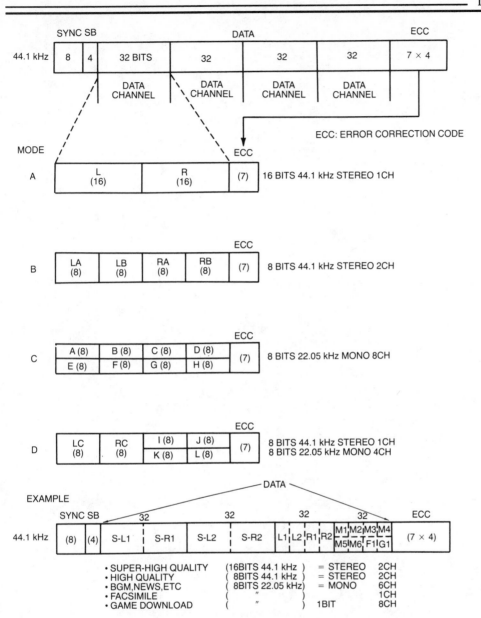

Fig. 5-42. CADA data format.

Videotex is the international name for systems which use telephone lines to connect home television sets to the computers in information centers to display, at the user's request, the information stored in the computer. This interactive system can access a wide variety of information including news, weather forecasts, educational programs, and information on leisure activities. Other services including reservations for travel and concert tickets, ordering of products, and banking would be handled over the network. This and other data transmission and storage media will be developed to satisfy our increasing appetite for information.

Chapter 6
Error Protection

Introduction

Error protection of digital data presents questions unique to the field of audio storage. With analog storage, there is no opportunity for error protection processing; if the recorded signal is disrupted or distorted then that signal is irrevocably damaged. With digital storage, the nature of binary data lends itself to recovery in the event of damage. When data is initially recorded, it may be pre-conditioned with error detection data which enables the reproduced data to be checked for error; if an error occurs, further signal recovery may be accomplished to correct the error, either absolutely or marginally, or the error may be concealed by synthesizing new data. In all cases, error protection makes the digital data storage methods much more robust. Strong error protection techniques relax the manufacturing tolerances for mass media such as the Compact Disc.

However, it must be noted that error protection is more than a unique opportunity, rather, it is an obligation. Because of high data densities in audio storage media, a minute defect in the medium or an introduced obstacle, such as a particle of dust, may cause the loss of hundreds or thousands of bits. Compared to absolute numerical data stored digitally, where a bad bit might mean the difference between adding or subtracting figures from a bank account, digital audio data is relatively forgiving of errors; our enjoyment of music is not necessarily ruined because of a few bad bits. However, error protection is mandatory for a digital audio system because larger errors are audible and these can easily occur because of the relatively harsh environment most audio media are subjected to. Error protection for digital audio is thus an opportunity to preserve data integrity, an opportunity not available with analog storage, and error protection is absolutely necessary to ensure the success of digital audio storage, because errors surely occur. With proper design engineering, digital audio storage systems, such as the Compact Disc system, can approach the computer industry standard which specifies an error rate of 10^{-12}, that is, less than one uncorrectable error in 10^{12} (one trillion) bits; however, much less stringent error performance is fully adequate for most audio applications.

6.1 Sources of Errors

High recording densities and unfriendly environment necessitate a sophisticated error protection scheme. Without such protection, it is doubtful whether digital audio recording would ever have become viable; indeed, the evolution of digital audio technology can be measured by the prerequisite advances in error protection technology. Errors can occur in every stage of the digital audio recording chain; however, the focus of error protection lies in the recorded media because it is errors contained in the media itself, or errors which are impressed on the media during use which are most severe and least subject to control. A summary of errors is given as follows:

I   System Errors
    1. aliasing
    2. quantization
    3. phase distortion
    4. nonlinearity
    5. noise
    6. aperture error
    7. inter-symbol interference
    8. transport irregularity (jitter)
    9. mistracking or misfocusing
    10. crosstalk

II   Media Errors
    A   Magnetic Media
        1. dust
        2. scratches
        3. fingerprints
        4. tape stretching or abuse
        5. impure oxide or binder
        6. irregular tape slitting
        7. drop-outs
        8. editing
    B   Optical Media
        1. dust
        2. scratches
        3. fingerprints
        4. pit asymmetry
        5. bubbles or defects in substrate
        6. coating defects
        7. drop-outs

6.1-1 Fundamental Errors • Fundamental types of errors occur in several guises in an audio digitization system. As we have seen, quantization is a process of approximation thus quantization noise itself is a type of error; with a large number of bits in a PCM system, we can reduce quantization noise to a non-objective small value. Similarly, aliasing could pose problems to the quality of the recorded signal; careful design of the input low-pass filter is required to minimize the effects of aliasing. The electrical components of the recording chain itself can contribute errors to the data; the low-pass filters, sample and hold circuit, and A/D and D/A converters, can all limit frequency response and signal-to-noise ratio, and add distortion, noise, and nonlinearity. However, by using high quality components, careful circuit design, and strict manufacturing procedures, these types of errors can be minimized. For example, a high quality A/D converter will exhibit satisfactory performance specifications, and careful construction will reduce the chance of induced noise and hum.

6.1-2 Media Errors • The most severe types of errors occur primarily in the recording medium itself; just as with analog technology, it is by far the weakest link in the recording chain. However, digital techniques offer the opportunity to detect and correct or otherwise negate the effects of many errors in the medium. Two primary types of potential errors are caused by transport jitter and tape or disc drop-outs.

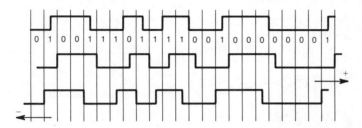

Fig. 6-1. Jitter timing error: Irregularities in the transport can cause timing errors in the output bit stream.

Jitter is any time-axis variation in the recording mechanism, with analog technology this is manifested as wow and flutter. In a digital system speed variations in the transport carrying the tape or disc past the pickup will cause the data rate to vary; the transport's speed can either slowly drift up and down about the proper speed, or fluctuate rapidly, as illustrated in Fig. 6-1. If the amount of variation is within the jitter margin, that is, if the proper value of the recorded waveform can still be recovered, then no error in the output signal will result. Servo control circuits can be used in which timing information is read from the medium and used to generate correct speed control signals for the medium's transport, as shown in Fig. 6-2. To correct even minute jitter

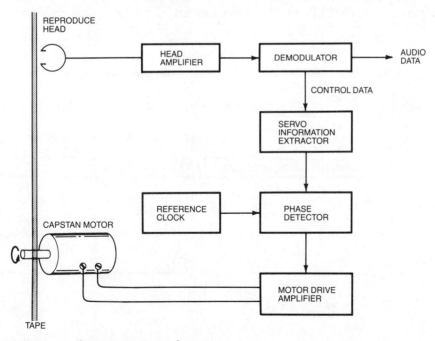

Fig. 6-2. Servo pulse extractor/control system.

errors, all digital systems contain an output buffer, a memory through which the data is clocked before it is output. Although the data rate into the buffer might vary because of jitter, the output will be constant because the data rate is clocked with an accurate crystal oscillator. In extreme cases, where the jitter causes the data rate's variation to exceed the buffer's length, and ability to correct the cause of the jitter itself, the transport or the servo system controlling the transport will require repair.

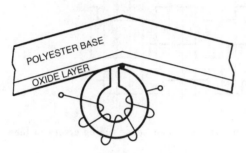

Fig. 6-3. **Spacing loss causing drop-out error.**

The most significant cause of data errors are drop-outs. In analog systems, a defect in the media causes a momentary drop in signal strength, hence the term "drop-out." With digital storage, an obstruction between tape and head, or defects in the medium can cause a similar error. Drop-outs can occur in any magnetic tape or optical disc media, and can be traced to two causes: a defect introduced during the use of the medium or a manufactured defect in the medium. During recording or playback, a particle could cause the tape to lift away from the head thus the recorded or playback signal strength would fail momentarily; this is known as spacing loss and is shown in Fig. 6-3. Magnetic tape and optical discs are manufactured under clean conditions; however, microscopic particles of dust and foreign particles can still enter into the manufacturing materials. This might produce an error sometimes called a drop-in;

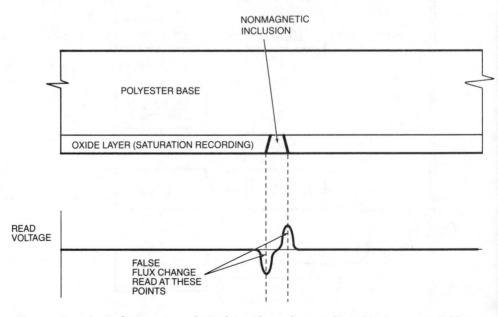

Fig. 6-4. **Drop-in. Inclusions or cracks in the oxide produce reading of a misrepresented bit.**

defects in the media cause signal transitions which are misrepresented as data, as shown in Fig. 6-4. Because of its familiarity, the term drop-out is commonly used to describe both error conditions.

In a digital audio recording, any loss of data, or invalid data, may provoke a click, or an explosive sound, as the D/A converter's output suddenly jumps to a new amplitude as it attempts to output missing or incorrect data. The severity of the error depends on the nature of the error; a bad bit in the least significant part of a PCM word might pass unnoticed whereas an invalid most significant bit would create a drastic change in amplitude.

Errors resulting from drop-outs occur in several modes thus classifications have been developed to better identify them. Drop-out errors which have no relation to each other are called random-bit errors, as shown in Fig. 6-5. They occur singly and can be easily corrected; they are more often found in optical discs, rather than magnetic media. A burst error is the most common kind of error found in magnetic media; it is a large error disrupting perhaps hundreds of bits and might be caused by a manufacturing defect or a dust particle, as shown in Fig. 6-6. Because the nature of errors depends on the medium, error protection techniques must be optimized for the application at hand. Thus, both magnetic and optical media errors must be considered separately. However, one or both of these errors may occur within a single medium, thus any error protection system must be designed to deal with both kinds of errors.

Several error conditions are encountered in digital magnetic media. As we have seen, drop-outs can occur during the use of a magnetic medium; because

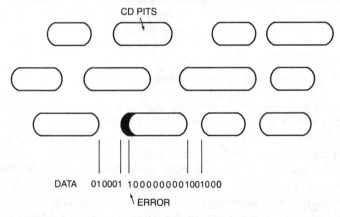

(A) Compact Disc random bit error caused by badly formed pit edge.

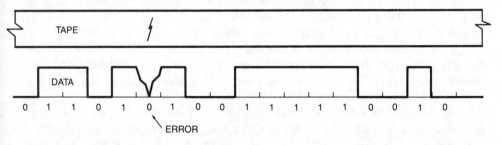

(B) Magnetic tape defect causing bit error.

Fig. 6-5. Random bit errors caused by a slight manufacturing defect cause single bit errors.

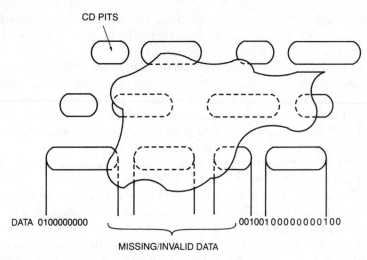

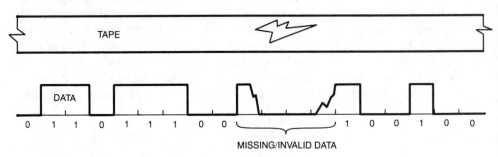

DATA 0100000000   001001 00000000100

MISSING/INVALID DATA

(A) Compact Disc burst error.

TAPE

DATA

0   1   1   0   1   1   1   0   0   1   0   0   1   0   0

MISSING/INVALID DATA

(B) Magnetic tape burst error.

Fig. 6-6. **Burst errors affect larger numbers of bits.**

of normal wear and tear, oxide particles will come loose from the backing and may become misplaced on the tape. A host of other foreign particles, such as dust and dirt, and oil from fingerprints, and tape guides causing scratches in the tape's oxide, all contribute to the number of drop-outs. A drop-out may occur at a fixed spot on the tape, or may travel to various spots on the tape as the errant particle moves, to create a phantom drop-out. Manufactured tape drop-outs usually occur more often at the beginning and end of a tape because of the slitting process in manufacture, due to the greater likelihood of variations in the cut width of the tape at the ends. Tape splicing causes errors in areas surrounding the edit point; error correction must provide a smooth output waveform to conceal the disrupted data as shown in Fig. 6-7.

Inter-symbol interference is caused when magnetically recorded bit patterns become indistinct; this results from a combination of problems in the recording circuitry, both in tape and heads. If the bandwidth of the recording system is insufficient, or the recording density too great, the recorded waveform is reduced in amplitude, and interference between adjacent waveforms creates a phenomenon called peak shift; reproduction of data becomes difficult and errors are generated. Peak shift is illustrated in Fig. 6-8. In a recording system with sufficient bandwidth the reproduced waveform follows the originally recorded one. With insufficient bandwidth the reproduced current wave-

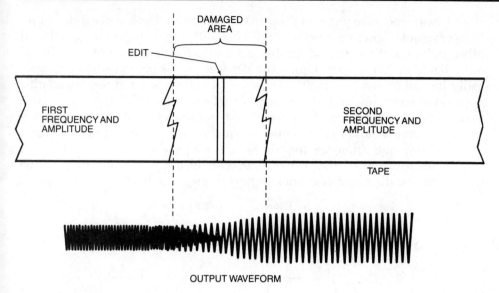

Fig. 6-7. **Splicing error and concealment.**

forms (dotted lines) mutually interfere. The resulting peak shift produces a broadened, incorrect waveform.

Noise is a cause of data errors; inadequate tape formulations, signal interference between channels called crosstalk occurring in wiring between channels and in tape heads, and problems attributed to the power supply may contribute to an increase in the noise floor which could degrade the signal quality and thus increase errors.

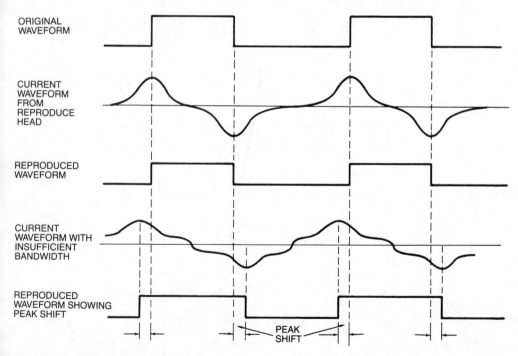

Fig. 6-8. **Inter-symbol interference.**

Optical media are prone to specific kinds of errors. Optical discs also suffer from drop-outs; however, they occur through different manufacturing and handling problems. When the master disc is manufactured, an incorrect amount of laser light, length of developing, or a defect in the photo-resist can create badly formed pits on the disc. Dust or scratches which occur during the plating or pressing processes, or pinholes or other defects in the reflective coating used to cover the disc can all create dropouts. As is the case with magnetic tape, stringent manufacturing conditions and quality control can prevent many of these drop-outs from leaving the factory.

During use, optical discs can be damaged; dust, dirt, and oil can be wiped clean from the disc; however, some scratches may interfere with the pickup's

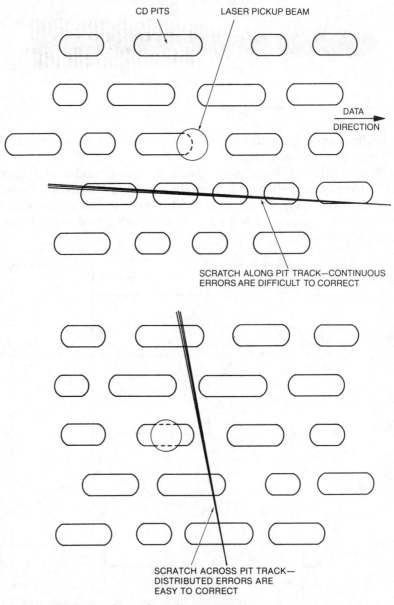

Fig. 6-9. Compact Disc scratches and resultant errors.

ability to read data from the disc. Whenever an optical disc, such as a Compact Disc, is cleaned, it should be done with a clean, soft cloth, and wiped radially, that is, across the width of the disc and not around its tracks; any scratches which result will be across many tracks and thus easier to correct whereas a single scratch along one data track could be impossible to correct because of the sustained consecutive loss of data, as shown in Fig. 6-9.

6.1-3 Objectives of Error Protection • Although a perfect error protection system is possible, in which every error is detected and corrected, it would create an unreasonably high data overhead because of the amount of redundant data required to accomplish it. Thus, an efficient error protection system should aim to provide a sufficiently low error rate after correction and concealment while minimizing the amount of redundant data, and data processing, required for successful operation. An error protection system is comprised of three operations; error detection in which redundant data permits data to be checked for validity, error correction in which data processing replaces bad bits with newly calculated valid bits, and in the event of large errors or insufficient data for correction, error concealment techniques may be employed to substitute approximately correct data for the bad data. In the worst case, when not even error concealment is possible, most digital audio systems choose to mute the output signal rather than let the output circuitry attempt to decode severely incorrect data, thus producing severely incorrect sounds.

6.2 Error Detection

All error detection and correction techniques are based on the redundancy of data. Information systems rely heavily on redundancy to achieve reliable communication; for example, spoken and written language contains redundancy. If an error-ridden telegraph message was received (e.g., AJL IS FIRGIVRN. PLEAOE COMW HOME.) the message could be recovered. In fact, Shannon estimated that 75 percent of written English is redundant.

Similarly, redundancy is required for reliable data communication. If a data value alone is generated, transmitted once, and received there is no absolute way to check its validity at the receiving end. We might examine the data word by word, and question, for example, a word which unexpectedly differs from its neighbors. With digital audio, in which the signal is usually relatively predictable, at least from one forty thousandth of a second to the next, such an algorithm might be reasonable; however, we could not absolutely detect errors, or begin to correct them, as shown in Fig. 6-10. Clearly additional information is required to reliably identify errors in received data; moreover, such infor-

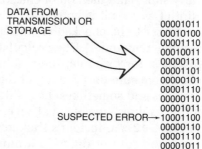

DATA FROM
TRANSMISSION OR
STORAGE

```
00001011
00010100
00001110
00010011
00000111
00001101
00000101
00001110
00000110
00001011
SUSPECTED ERROR→10001100
00000110
00001110
00001011
```

Fig. 6-10. **Without special encoding, errors cannot be reliably detected in received data.**

mation must be essentially redundant thus it must originate from the same point as the original data; it is thus subject to the same error-creating conditions as the data itself. The task of error detection is to properly code transmitted or stored information such that when data is lost or made invalid the coded information will correctly alert us to that fact.

6.2-1 Data Repetition • In an effort to detect errors, the original message could simply be repeated; however, this would quickly multiply the amount of data required to be stored. For example, each data word could be repeated twice, but given a conflict between repeated words, it would be impossible to identify the correct version. If each word were repeated three times the probability would be that the two in agreement were correct while the differing third was in error. Yet all three words could agree, and yet all be in error, unknown to us. Given enough repetition, the probability of correctly detecting an error would be high; however, as we have noted, the data overhead would be enormous thus more efficient methods are required. The limitations of data repetition are shown in Fig. 6-11. Data words could be transmitted or stored, once, twice, or three times but positive error detection is not achieved. In Fig. 6-11D are the three suspected words actually correct, or has the error been repeated three times? In any case, the data load is multiplied. Also, the increased data load may introduce errors, as shown.

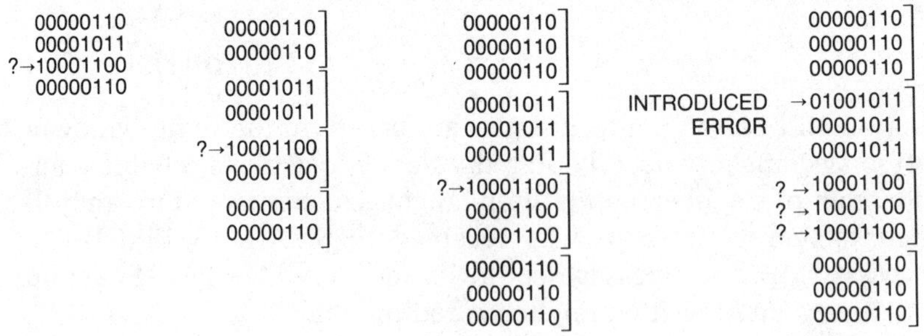

(A) One. (B) Two. (C) Three. (D) Introduced errors.

Fig. 6-11. **Data repetition as error detection.**

6.2-2 One-Bit Parity • A more efficient technique for error detection employs shorthand systems in which coded redundant data is used to check for errors. Parity and Cyclic Redundancy Check Codes (CRCC) are two such methods. One early shorthand method of checking for errors was devised by the Arabs in the Ninth Century; it is known as "casting out 9s." In this technique, a residue for a number to be checked is created by dividing the number by nine, leaving a residue (remainder). Then calculations may be checked for errors by comparing residues: for example, the residue of the sum (or product) of two numbers equals the sum (or product) of the residues. It is important to always compare residues and sometimes the residue of a sum or product residue must be taken. If the residues are not equal, an error has occurred in the calculation, as shown in Fig. 6-12. An insider's trick makes the shorthand method even more shorthand: the sum of digits in a number always has the same 9s residue as the number itself. The technique of casting out 9s can be used to cast out any

CASTING OUT 9s:

$240 + 578 \stackrel{?}{=} 818 \rightarrow (2 + 4 + 0 = 6)$

$(5 + 7 + 8 = 20, 2 + 0 = 2)$

$(8 + 1 + 8 = 17, 1 + 7 = 8)$

$6 + \quad 2 \stackrel{\checkmark}{=} \quad 8$ CASTING OUT 9s SUM
AGREES THUS NO ERROR

$227 \times 67 \stackrel{?}{=} 15209$

$2 \times \quad 4 \stackrel{\checkmark}{=} \quad 8$ CASTING OUT 9s PRODUCT
AGREES THUS NO ERROR

$154 \times 95 \stackrel{?}{=} 14613$

$1 \times \quad 5 \neq \quad 6$ CASTING OUT 9s PRODUCT
DOES NOT AGREE.
CALCULATION IS IN ERROR

CASTING OUT 2s:

SUM OF 11001011 IS 5, WHICH IS ODD,
CAST OUT 2s TO GET 1 AND APPEND TO
WORD: 110010111. IN THIS WAY, THE NUMBER
OF 1s IS ALWAYS EVEN.

Fig. 6-12. **Casting out of 9s and 2s.**

number, and in fact it forms the basis for a principle method of binary error detection, parity.

Given a binary number, a residue bit may be formed by casting out 2s. This extra bit is formed when the word is transmitted or stored, and is carried along with the data word. This extra bit, known as one-bit parity, permits error detection, but not correction. Rather than rely on casting out of 2s, a more efficient algorithm may be employed. The parity bit is formed with a simple rule—if the number of 1s in the data word is even (or zero), the parity bit is made 0; if the number of 1s in the word is odd, the parity bit is a 1. In other words, the data bits are added together with modulo 2 addition, as shown in Fig. 6-13. Thus, an 8 bit data word, made into a 9 bit word with an even parity bit, will always have an even number of 1s (or none). This parity scheme results in even parity. By reversing the parity bit, odd parity results; both schemes are functionally identical.

At playback, the validity of the received data word is tested by the parity bit, that is, the received data bits are added together to calculate parity of received data; if the received parity bit and the calculated parity bit are in conflict, then an error has occurred. Probability dictates that the error is in the data word, rather than the parity bit itself; however, the reverse could be true, and the parity bit error could be detected. Also, there could be more than one error in the received word which could defeat detection altogether, as shown in Fig. 6-14. In digital audio storage media, particularly magnetic media, errors tend to occur as burst errors, thus many errors would occur within each word and one-bit parity would not provide reliable detection. Thus, a one bit parity check code error detection system by itself is not suitable for digital audio storage or transmission.

6.2-3 Cyclic Redundancy Check Code • The CRCC is an error detection method preferred to one-bit parity in audio applications because of its increased ability to detect errors caused by burst errors in the recording media. The CRCC is a cyclic block code which generates a parity check word. For example, the bits of a data word could be added together to form a sum of the bits; this would be the parity check word. For instance, the six binary 1s in 1011011010 would be

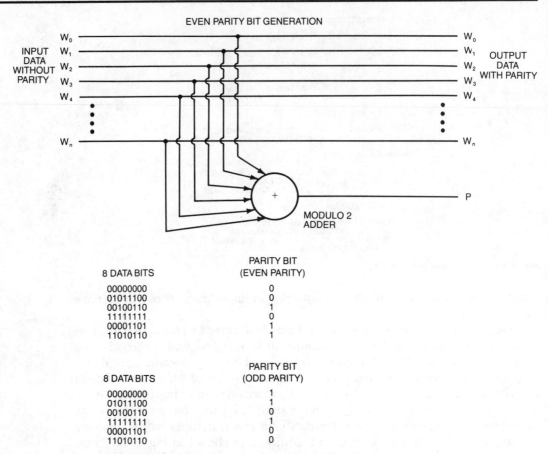

Fig. 6-13. Parity is formed through examination of data bits.

EVEN PARITY BIT GENERATION

8 DATA BITS	PARITY BIT (EVEN PARITY)
00000000	0
01011100	0
00100110	1
11111111	0
00001101	1
11010110	1

8 DATA BITS	PARITY BIT (ODD PARITY)
00000000	1
01011100	1
00100110	0
11111111	1
00001101	0
11010110	0

added together to form binary 0110 (6 in base 10), and that check word would be appended to the data word to form the code word for transmission or storage. As with a one-bit parity, any disagreement between the received checksum and that formed from the received data would indicate with high probability that an error has occurred.

TRANSMITTED WORD		RECEIVED WORD		PARITY CALCULATED FROM RECEIVED DATA WORD	
DATA	PARITY	DATA	PARITY		
00011001	1	00001001	1	0	ERROR DETECTED
10101011	1	11001011	1	1	ERRORS NOT DETECTED
01110100	0	01110100	1	0	PARITY ERROR DETECTED
01101011	1	00000011	0	0	ERRORS NOT DETECTED

Fig. 6-14. One-bit parity error detection.

The CRCC works similarly, but with a more sophisticated calculation; simply stated, each data block is divided by an arbitrary and constant number, and the remainder of the division is appended to the data block and stored or transmitted. Upon reproduction, the division is repeated, and the remainder is compared with that appended to the data word, as shown in Fig. 6-15.

A more detailed examination of the CRCC scheme for a k bit block and n − k bit detection block shows the intermediate steps in coding and decoding.

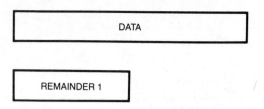

(A) Original block of data is divided to produce a remainder.

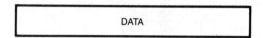

(B) The two words are transmitted or stored together.

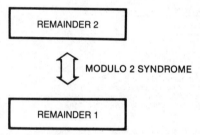

(C) The received data word is again divided to produce a remainder.

(D) This is compared to the received remainder; any difference indicates an error.

Fig. 6-15. CRCC in simplified form.

The original k bit data block **m** is multiplied by X^{n-k} then divided by the generation polynomial **g** to form the quotient **q** and remainder **r**. The transmission polynomial **v** is formed from the original message **m** and the remainder **r**; thus it is a multiple of the generation polynomial **g**. The transmission polynomial **v** is then transmitted or stored. The received data **u** undergoes error detection by calculating a syndrome **s** with modulo 2 addition of received parity bits and parity newly calculated from the received message. A zero syndrome shows an error-free condition. A nonzero syndrome denotes an error. Error correction may be accomplished by forming an error pattern, which is the difference between the received data and the original data to be recovered. This is mathematically possible because the error polynomial **e** divided by the original generation polynomial produces the syndrome as a remainder. Thus the syndrome may be used to form the error pattern, and hence recover the original data. It is important to choose **g** such that the error patterns in the error polynomial **e** are not evenly divisible by **g**. The CRCC encoding and decoding steps are shown in Fig. 6-16. A numerical example of encoding of a cyclic code is shown in Fig. 6-17. A shift register implementation of an encoder is shown in Fig. 6-18.

It can be readily seen that the larger the data block chosen, the less redundancy results yet mathematical analysis shows that the same error detection

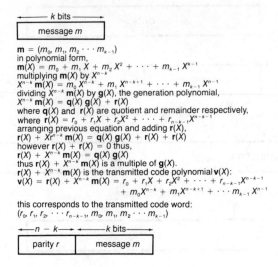

(A) CRCC encoding.

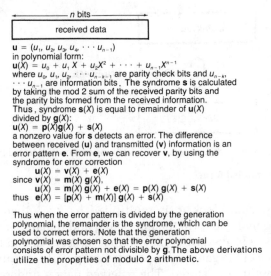

(B) CRCC decoding and syndrome calculation.

Fig. 6-16. CRCC encoding and decoding steps.

capability remains. However, if random or short duration burst errors tend to occur frequently, then the integrity of the detection is decreased and shorter blocks may be necessitated.

The extent of error detectability of a CRCC scheme may be summarized. Given a k bit data word with m (m = n−k) bits of CRCC, a code word of n bits is formed, and the following is true:

1. Burst errors less than or equal to m bits are always detectable.
2. Misdetection probability of burst errors longer than m + 1 bits is 2^{-m+1}.
3. Misdetection probability of burst errors longer than m + 1 bits is 2^{-m}.
   (Items 1–3 are not affected by the length n of the code word.)
4. Random errors up to 3 consecutive bits long can be detected.

Given a generation polynomial $g(X) = 1 + X^2 + X^3$ and a message $m = (1001)$ to be encoded. Message polynomial $m(X) = 1 + X^3$. Multiplying by X^{n-k}, $X^3m(X) = X^3 + X^6$. Division by $g(X)$ is accomplished:

$$
\begin{array}{r}
X^3 + X^2 + X + 1 \\
X^3 + X^2 + 1 \overline{\smash{\big)}\,X^6 \qquad\qquad X^3} \\
\underline{X^6 + X^5 \qquad + X^3} \\
X^5 \\
\underline{X^5 + X^4 \qquad\quad + X^2} \\
X^4 \qquad\qquad + X^2 \\
\underline{X^4 + X^3 \qquad\quad + X} \\
X^3 + X^2 + X \\
\underline{X^3 + X^2 \qquad\quad + 1} \\
X + 1 \quad \text{remainder} = r(X)
\end{array}
$$

The code word polynomial $v(X) = r(X) + X^3 m(X)$.
$\qquad\qquad\qquad = 1 + X + X^3 + X^6$.
The code word is thus $(\underline{110}\underline{1001})$ which corresponds to $(1, X, X^{2^0}, X^3, X^{4^0}, X^{5^0}, X^6)$.
$\qquad\qquad$ parity message

Fig. 6-17. **A numerical example of encoding of a cyclic code.**

It is ultimately the medium itself which determines the design of the CRCC, and the rest of the error protection system. Magnetic tape, for example, might call for longer CRCC blocks, while optical discs might require shorter blocks. The power of the error correction processing following the CRCC also influences how accurate the CRCC detection must be. The CRCC is typically used as an error pointer to identify the number and extent of errors prior to other error correction processing.

6.3 Error Correction

With the use of redundant data it is possible to correct errors which occur during transmission or storage of digital audio data. In the simplest case, data is simply repeated. For example, instead of writing only one data track to recorded tape, two tracks of identical data could be written. The first track would normally be used for playback but if an error is detected through parity or other means, then data is taken from the second track. To help ensure that simultaneously bad data occurs rarely, redundant samples could be staggered with respect to each other in time. Furthermore, data bits could be interleaved to help distribute errors. While such a dual-track scheme is workable, it is inefficient because it doubles the amount of data to be stored. A more enlightened approach is that of error correcting codes; they can achieve more reliable transmission or storage with less redundancy. In the same way that coded redundant data in the form of parity check bits is employed for error detection, redundant data is used to form codes for error correction. Digital audio is encoded with separate detection and correction schemes; upon decoding, errors are identified by the detection decoder, and corrected by the correction decoder. Coded redundant data is the essence of all correction codes; however, there are many types of codes, different in their design and function.

6.3-1 Error Correction Codes • The field of error correction codes is a highly mathematical one; many types of codes, resulting from many differing theories concerning the best approach to developing codes for different applications have been proposed. In general, two approaches, shown in Fig. 6-19, have been developed, block codes using algebraic methods and convolutional codes using probabilistic schemes. A block code generated by an encoder is formed solely from the message currently in the block. In a convolutional code the generated

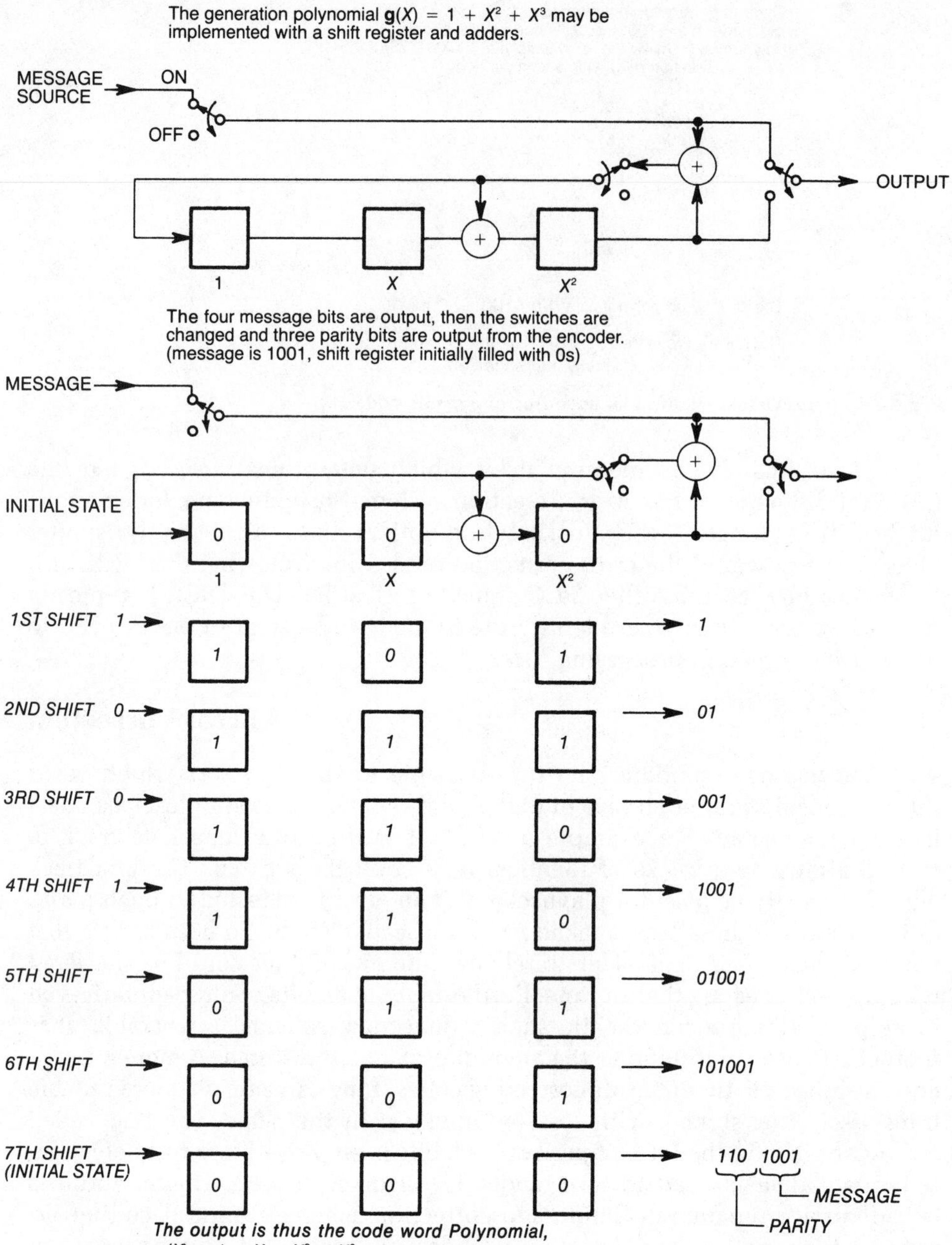
The generation polynomial g(X) = 1 + X² + X³ may be implemented with a shift register and adders.
MESSAGE SOURCE
ON
OFF
OUTPUT
1
X
X²
The four message bits are output, then the switches are changed and three parity bits are output from the encoder. (message is 1001, shift register initially filled with 0s)
MESSAGE
INITIAL STATE
0 0 0
1ST SHIFT 1 → 1 0 1 → 1
2ND SHIFT 0 → 1 1 1 → 01
3RD SHIFT 0 → 1 1 0 → 001
4TH SHIFT 1 → 1 1 0 → 1001
5TH SHIFT → 0 1 1 → 01001
6TH SHIFT → 0 0 1 → 101001
7TH SHIFT (INITIAL STATE) → 0 0 0 → 110 1001
MESSAGE
PARITY
The output is thus the code word Polynomial, v(X) = 1 + X + X³ + X⁶

Fig. 6-18. **An implementation of a cyclic encoder.**

coded message is formed from the message present in the encoder at that time including previous messages. In addition to a block versus convolutional distinction, error correction codes may be considered as linear versus nonlinear and word versus bit. In general, for digital audio, linear and word methods have been adopted with block code, or block code in a convolutional structure, known as a cross-interleave code. Such codes are used in the DASH, and CD formats.

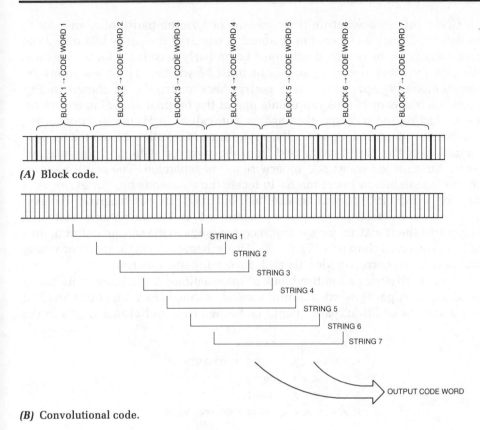

(A) Block code.

(B) Convolutional code.

Fig. 6-19. **Block versus convolutional codes.**

6.3-2 Block Code • With increased redundancy, coded in an intelligent fashion, error detection methods, such as parity, may be extended to provide error correction. The correction encoding may be accomplished by generating coded parity bits for a data word, or by assembling a number of data words to form a block, and attaching a parity word to the block. During decoding an error correction algorithm may be used to form a syndrome word which checks for errors and given enough redundancy, correct them. An extension of the technique employs both individual word and block parity simultaneously. Such schemes have proved to be effective against errors encountered by digital audio media. Error correction is enhanced by interleaving of individual bits in consecutive words, and word interleaving.

Cyclic codes, such as CRCC, are a subclass of linear block codes; they may be used to achieve error correction capability. Special block codes, known as Hamming codes create syndromes which point to the actual location of the error. Multiple parity bits are formed for each data word, with unique encoding. For example, three parity check bits (4, 5, and 6) might be added to a four bit data word (0, 1, 2, and 3); 7 bits are transmitted. Suppose that the three parity bits are uniquely defined such that parity bit 4 is formed from modulo 2 addition of data bits 1, 2, and 3, parity bit 5 is formed from data bits 0, 2, and 3, and parity bit 6 is formed from data bits 0,1, and 3, as shown in Fig. 6-20A. Thus, for example, the data word 1100, appended with the parity bits 110 would be transmitted as the seven bit code word 1100110.

Given this somewhat unlikely scheme of forming parity bits, an error in the data word may be located by examining which of the parity bits reflects an error. Of course, the received data must be properly decoded prior to checking. Thus, parity check decoding equations must be written. These equations are computationally represented as a parity-check matrix H, as shown in Fig. 6-20B. Each row of H thus represents one of the original encoding equations, and by testing the received data against the values in H, the location of the error may be identified. Specifically, a syndrome is calculated from the modulo 2 addition of the parity calculated from the received data, and the received parity. An error generates a 1, otherwise a 0 is generated. The resulting error pattern is matched in the H matrix to locate the erroneous bit. For example, if the code word 1100110 was transmitted, but 1000110 was received, the syndromes would detect the error and generate a 1, 0, 1 error pattern. Matching this against the H matrix, we see that it corresponds to the second column, thus bit 1 is in error, as shown in Fig. 6-20C. This scheme is a single error correcting code thus it can correctly identify and correct any one-bit error.

Block parity uses an entire block of information for the basis of its parity calculations. A parity word is simultaneously formed from the entire block of data words. In addition, parity could be formed from individual words in the

$$X_0, X_1, X_2, X_3 \qquad \text{DATA BITS}$$

$$X_4 = X_1 + X_2 + X_3 \qquad \text{(MOD 2) PARITY CHECK BITS}$$

$$X_5 = X_0 + X_2 + X_3 \qquad \text{(MOD 2)}$$

$$X_6 = X_0 + X_1 + X_3 \qquad \text{(MOD 2)}$$

$$X_0, X_1, X_2, X_3, X_4, X_5, X_6 \quad \text{TRANSMITTED CODE WORD}$$

(A) Parity bits are formed.

$$\begin{array}{llll} X_1 + X_2 + X_3 + X_4 & = 0 & \text{DECODING} \\ X_0 \quad + X_2 + X_3 \quad + X_5 & = 0 & \text{ALGORITHM} \\ X_0 + X_1 \quad + X_3 \quad + X_6 = 0 \end{array}$$

$$\begin{bmatrix} 0 & 1 & 1 & 1 & 1 & 0 & 0 \\ 1 & 0 & 1 & 1 & 0 & 1 & 0 \\ 1 & 1 & 0 & 1 & 0 & 0 & 1 \end{bmatrix} = H \quad \begin{array}{l}\text{PARITY-CHECK} \\ \text{MATRIX}\end{array}$$

(B) Parity check matrix.

EXAMPLE: 1100110   TRANSMITTED WORD
             1000110   RECEIVED WORD

$$P_4 = X_1 + X_2 + X_3 = 0 + 0 + 0 = 0 \qquad \text{PARITY OF RECEIVED DATA}$$
$$P_5 = X_0 + X_2 + X_3 = 1 + 0 + 0 = 1$$
$$P_6 = X_0 + X_1 + X_3 = 1 + 0 + 0 = 1$$

SYNDROMES ARE CALCULATED WITH MOD 2 ADDITION OF
PARITY OF RECEIVED DATA AND RECEIVED PARITY BITS:

$$\begin{array}{lll} P_4 = 0, \quad X_4 = 1 & 0 + 1 = 1 & \text{(ERROR)} \\ P_5 = 1, \quad X_5 = 1 & 1 + 1 = 0 & \text{(CORRECT)} \\ P_6 = 1, \quad X_6 = 0 & 1 + 0 = 1 & \text{(ERROR)} \end{array}$$

THE RESULTING SYNDROMES FORM THE ERROR PATTERN $\begin{bmatrix} 1 \\ 0 \\ 1 \end{bmatrix}$ WHICH

CORRESPONDS TO THE SECOND COLUMN OF H

$$H = \begin{bmatrix} 0 & 1 & 1 & 1 & 1 & 0 & 0 \\ 1 & 0 & 1 & 1 & 0 & 1 & 0 \\ 1 & 1 & 0 & 1 & 0 & 0 & 1 \end{bmatrix} \qquad \text{THUS BIT } X_1 \text{ IS IN ERROR.}$$

(C) Single error correction using syndromes.

Fig. 6-20. **Hamming error correction code.**

block, using one-bit parity or a cyclic code. In this way, greater redundancy is achieved, and correction is improved. For example, CRCC could be used to detect an error, then block parity used to correct the error. As an example, consider the expense account for a motorcycle trip, as shown in Fig. 6-21A. Twenty figures are recorded, but eight of them are redundant. Moreover, because of the way in which the redundancy is created, an incorrect entry can be detected by comparing the row and column totals; any single error will cause one row and one column total to be in error, thus the erroneous data can be traced to the intersection of that row and column. Do the numbers tally in Fig. 6-21B? If not, can you correct the error?

(A) Correct.

	Thursday	Friday	Saturday	Sunday	total
motels	$14.50	$18.75	$12.00	$12.00	$ 57.25
gasoline	$20.25	$15.00	$18.50	$14.00	$ 67.75
beer	$ 8.50	$ 6.25	$ 7.50	$ 9.00	$ 31.25
total	$43.25	$40.00	$38.00	$35.00	$156.25

(B) Correct?

	Thursday	Friday	Saturday	Sunday	total
motels	$18.00	$16.50	$25.00	$16.50	$ 76.00
gasoline	$20.50	$17.50	$10.00	$ 7.50	$ 55.50
beer	$ 5.50	$ 6.50	$ 6.00	$ 8.00	$ 29.00
total	$44.00	$40.50	$44.00	$32.00	$160.50

Fig. 6-21. **Motorcycle trip expense account.**

In similar fashion, a binary message might be consolidated into a block, with row and column parity. For example, a 12 bit message might be arranged in a 3 by 4 matrix, as shown in Fig. 6-22. Added to each row and column is a parity bit, as well as a bit at the intersection of the row and column which simultaneously provides parity for both. The information in the complete matrix could be transmitted or stored row by row or column by column. At the receiving end, the data is checked for correct parity; any single error can be corrected. Any double errors can be detected, but not corrected. Larger numbers of errors could result in misdetection or miscorrection. Each of the transmitted bits is called a symbol; thus the code word in this example has 20 symbols (n = 20), of which 12 (k = 12) are information. Codes are referred to

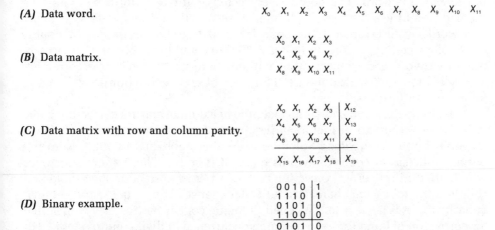

(A) Data word.

$$X_0 \quad X_1 \quad X_2 \quad X_3 \quad X_4 \quad X_5 \quad X_6 \quad X_7 \quad X_8 \quad X_9 \quad X_{10} \quad X_{11}$$

(B) Data matrix.

$$\begin{matrix} X_0 & X_1 & X_2 & X_3 \\ X_4 & X_5 & X_6 & X_7 \\ X_8 & X_9 & X_{10} & X_{11} \end{matrix}$$

(C) Data matrix with row and column parity.

$$\begin{array}{cccc|c} X_0 & X_1 & X_2 & X_3 & X_{12} \\ X_4 & X_5 & X_6 & X_7 & X_{13} \\ X_8 & X_9 & X_{10} & X_{11} & X_{14} \\ \hline X_{15} & X_{16} & X_{17} & X_{18} & X_{19} \end{array}$$

(D) Binary example.

$$\begin{array}{cccc|c} 0 & 0 & 1 & 0 & 1 \\ 1 & 1 & 1 & 0 & 1 \\ 0 & 1 & 0 & 1 & 0 \\ 1 & 1 & 0 & 0 & 0 \\ \hline 0 & 1 & 0 & 1 & 0 \end{array}$$

Fig. 6-22. **Message arranged in 3 by 4 matrix.**

as (n,k) codes, this is a (20, 12) code. The source rate is $R = k/n = 12/20$. There are $n-k = 20-12 = 8$ check symbols. The minimum distance of a code determines its error correctability; a minimum distance of 2 gives single error detection, 3 gives double error detection or single error correction, 4 gives triple error detection or double error detection plus single error correction, 5 gives double error correction, etc.

Block correction codes employ many methods to generate the transmitted code word and its parity; however, they are fundamentally identical in that only information from the block itself is used to generate the code. The extent of the correction capabilities of block correction codes may be simply illustrated by a decimal number example. Given a block of six data words, a seventh parity word can be calculated by adding the six data words. To check for an error, a syndrome is created by comparing (subtracting in the example) the parity (sum) of the received data with the received parity value. If the result is zero, then most probably no error has occurred as shown in Fig. 6-23A. A nonzero result would indicate that an error has probably occurred, as shown in Fig. 6-23B.

If one data word is missing, a condition called a single erasure, the syndrome would indicate that, furthermore the missing value could be obtained from the syndrome, as shown in Fig. 6-23C. If a data word is erroneous, the nonzero syndrome would indicate that; however, without further information the correct value cannot be calculated, as shown in Fig. 6-23D. If CRCC or one-bit parity had been used, it would point out the erroneous word, and the correct value could be calculated using the syndrome, as shown in Fig. 6-23E.

Even if CRCC, or one-bit parity, was itself in error and falsely created an error pointer, the syndrome would yield the correct result, as shown in Fig. 6-23F. Such a block correction code is capable of detecting a one word error, or making one erasure correction, or correcting one error with pointer. As we have seen, its ability to correct is dependent on the detection ability of pointers. When used for digital audio, the parity word is formed by modulo 2 addition and CRCC is used for error pointers. A binary example of a correction scheme with bit and block parity is shown in Fig. 6-24. Errors detected through one-bit parity are corrected with block parity.

For enhanced performance, two parity words could be formed for the data block. Any one word error could be corrected, if any two words were missing the code could use the two syndromes to supply the missing data, and if any two words were erroneous the code could detect this; examples of double parity block coding are shown in Fig. 6-25. This double parity code can accomplish two word erasure correction without pointers, and its error correction ability is increased with pointers. This type of error correction is well suited for audio applications.

The Reed-Solomon code is an example of a double erasure correction code; it is highly successful in digital audio applications when coupled with CRCC for error pointers. The Reed-Solomon code, first published in 1960, is an important subclass of codes known as q-ary BCH codes. These are cyclic codes which are multiple-error correcting codes. A t-error correcting Reed-Solomon code has the following characteristics: Block length of $n = q-1$, and number of parity digits is $n-k = 2t$. The Reed-Solomon code is particularly well suited for correction of burst errors, a common occurrence in digital audio media. For example, a double erasure correction Reed-Solomon code (minimum distance

ORIGINAL DATA WORDS AND PARITY

W_1 10
W_2 30
W_3 20
W_4 25
W_5 30
W_6 15
P 130 = $W_1 + W_2 + W_3 + W_4 + W_5 + W_6$

(A) Block correction code showing no error condition.

RECEIVED DATA WORDS AND PARITY

W_1 10
W_2 30
W_3 20
W_4 25
W_5 30
W_6 15
P 130

SYNDROME $S = W_1 + W_2 + W_3 + W_4 + W_5 + W_6 - P$
$= 10 + 30 + 20 + 25 + 30 + 15 - 130 = 0$

THUS NO ERROR IS INDICATED

RECEIVED DATA WORDS AND PARITY

W_1 10
W_2 30
W_3 10
W_4 25
W_5 30
W_6 15
P 130

(B) Block correction code showing error condition.

SYNDROME $S = 10 + 30 + 10 + 25 + 30 + 15 - 130 = -10$

THUS AN ERROR IS INDICATED

RECEIVED DATA AND PARITY WORD

W_1 10
W_2 30
W_3 —
W_4 25
W_5 30
W_6 15
P 130

(C) Block correction code showing single erasure correction.

SYNDROME $S = 10 + 30 + 0 + 25 + 30 + 15 - 130 = -20$

ERASURE CORRECTION: $W_3 = W_3' - S$
$= 0 - (-20) = 20$

RECEIVED DATA AND PARITY WORD

W_1 10
W_2 30
W_3 20
W_4 15
W_5 30
W_6 15
P 130

(D) Block correction code showing uncorrectable error correction.

SYNDROME $S = 10 + 30 + 20 + 15 + 30 + 15 - 130 = -10$

ERROR CORRECTION: NOT POSSIBLE

RECEIVED DATA AND PARITY WORD

W_1 10
W_2 30
W_3 20 ⟵ CRCC ERROR POINTER
W_4 15
W_5 30
W_6 15
P 130

(E) Block correction code showing correction with pointer.

SYNDROME $S = 10 + 30 + 20 + 15 + 30 + 15 - 130 = -10$

ERROR CORRECTION: $W_4 = W_4' - S$
$= 15 - (-10)$
$= 25$

RECEIVED DATA AND PARITY WORD

W_1 10
W_2 30
W_3 20
W_4 25 ⟵ FALSE ERROR POINTER
W_5 30
W_6 15
P 130

(F) Block correction code showing false pointer.

SYNDROME $S = 10 + 30 + 20 + 25 + 30 + 15 - 130 = 0$

ERROR CORRECTION: $W_5 = W_5' - S$
$= 30 - 0$
$= 30$

Fig. 6-23. **Examples of single parity block coding.**

```
TRANSMITTED          TRANSMITTED
DATA BLOCK           SINGLE BIT PARITY

00010111                   0
01101010                   0
10010111                   1
11010110                   1
00111100 ─────────────────────────── TRANSMITTED PARITY WORD


RECEIVED             RECEIVED
DATA BLOCK           PARITY BIT

00010111                   0
01101010                   0
11100100                   1
11010110                   1
00111100 ─────────────────────────── RECEIVED PARITY WORD
01110011 ─────────────────────────── PARITY WORD CALCULATED FROM
                                      RECEIVED DATA AND PARITY WORD


PARITY BITS CALCULATED
ON RECEIVED DATA BLOCK

        0
        0
        0 ──────── INDICATES ERROR IN WORD 3
        1


    01110011 ──── CALCULATED PARITY WORD
 +  11100100 ──── INCORRECT WORD 3
    10010111 ──── CORRECTED WORD 3
```

Fig. 6-24. **Block parity.**

of 3) with CRCC error pointers has been adopted for one manufacturer's stationary head format. A quadruple-erasure (double error) correction Reed-Solomon code (minimum distance of 5), known as the Cross-Interleave Reed-Solomon Code (CIRC) has been adopted for the Compact Disc.

6.3-3 Convolutional Code • Convolutional codes, sometimes called recurrent codes, differ from block codes in the way data is grouped for coding. Instead of dividing the message data into large blocks of k digits and generating a block of n code digits, convolutional codes do not partition data into large blocks. Instead, message digits k are taken a few at a time and used to generate coded digits n, formed not only from those k message digits, but from many previous k digits as well, saved in delay memories. Such a code is called a (n,k) convolutional code. It uses $(N-1)$ message blocks with k digits. It has constraint length N blocks (or nN digits) equal to $n(m+1)$ where m is the number of delays. Its rate R is k/n. Parameters k and n typically are small integers.

As in the case of linear block codes, encoding is accomplished, and code words transmitted or stored, then upon retrieval the code words are checked for errors using syndromes. Shift registers are often used to implement the delay memories required in the encoder and decoder. The amount of delay determines the code's constraint length, which is analogous to the block length of a block code. An example of a convolutional encoder as demonstrated by Hagelbarger is shown in Fig. 6-26. There are six delays, thus the constraint length is 14. The other parameters are $q = 2$, $R = \frac{1}{2}$, $k = 1$, $n = 2$, and polynomial $H(y,z) = y(1 + z^2 + z^5 + z^6)$. As can be seen from the diagram, message data is continually circulating through the encoder, and many previous bits affect the current coded output.

Another example of a convolutional encoder demonstrated by Viterbi is shown in Fig. 6-27A. The upper code is formed from the input data with the polynomial $(1 + x + x^2)$, and the lower with $(1 + x^2)$. The data sequence enters the circuit from the left and is shifted to the right one bit at a time, the two sequences generated from the original sequence with modulo 2 addition are multiplexed to again form a single coded data stream. The resultant code has a memory of 2 because in addition to the current input bit, it also acts on the preceding two bits. For every input bit there are two output bits hence the code's rate is one-half. The constraint length of this code is $k = 3$.

A convolutional code can be understood by a tree diagram, such as shown in Fig. 6-27B, as demonstrated by Viterbi. This represents the first five sequences for an infinite tree with nodes spaced n digits apart and 2^k branches leaving each node. Each branch is an n digit code block which corresponds to a specific k digit message block. Any code word sequence is represented as a path through the tree. For example, given the code word from the previous encoder, all possible encoding sequences for the first five sequential input bits are shown in the tree; if the input bit is a 0, the code symbol is obtained by going up to the next tree branch, and by going down if the input is a 1. The input message thus dictates the path through the tree, each input digit giving one instruction. The sequence of selections at the nodes forms the output code word. For example, the output path corresponding to the input 11100 would be 1101100111. Upon playback, the data is sequentially decoded and errors can be detected and recovered by comparing all possible transmitted sequences to those actually received. The received sequence is compared to all possible transmitted sequences branch by branch. The decoding path through the tree is guided by the algorithm to find the most likely transmitted sequence which gave rise to the received sequence.

Another convolutional code suggested by Doi is shown in Fig. 6-28; the encoder uses four delays, each delaying for a duration of one word. Check words are generated after every four data words and each check word has encoded information derived from the previous eight data words. The decoder shown in Fig. 6-29 uses four delay units, and four more delays to form syndromes, plus a matrix of 16 delays; any single error can be corrected using a simple algorithm in which the syndromes are subtracted from the data words, as shown in Table 6-1. This convolutional code has the same redundancy overhead as a four word block code, however, the convolutional code has better error correction capabilities because each syndrome contains information derived from a greater number of data words. One disadvantage of convolutional codes is error propagation; any error which cannot be fully corrected will generate syndromes reflecting this error and this could introduce errors in subsequent data.

Table 6-1. One-Word Correction Algorithm Table

Erroneous Word	Nonzero Syndrome	Correction
W_0'	S_0 and S_4	$W_0 = W_0' - S_4$
W_1'	S_0 and S_8	$W_1 = W_1' - S_8$
W_2'	S_0 and S_{12}	$W_2 = W_2' - S_{12}$
W_3'	S_0 and S_{16}	$W_3 = W_3' - S_{16}$

RECEIVED DATA AND TWO PARITY WORDS

$$\begin{array}{ll}
W_1 & 10 \\
W_2 & 30 \\
W_3 & 20 \\
W_4 & 25 \\
W_5 & 30 \\
W_6 & 15 \\
P & 130 = W_1 + W_2 + W_3 + W_4 + W_5 + W_6 \\
Q & 440 = 6W_1 + 5W_2 + 4W_3 + 3W_4 + 2W_5 + W_6
\end{array}$$

SYNDROME

$$S_1 = W_1 + W_2 + W_3 + W_4 + W_5 + W_6 - P = 10 + 30 + 20 + 25 + 30 + 15 - 130 = 0$$

$$S_2 = 6W_1 + 5W_2 + 4W_3 + 3W_4 + 2W_5 + W_6 - Q = 60 + 150 + 80 + 75 + 60 + 15 - 440 = 0$$

(A) Block correction code with double parity showing no error condition.

RECEIVED DATA AND PARITY WORDS

$$\begin{array}{ll}
W_1 & 10 \\
W_2 & 20 \leftarrow \text{ERROR POINTER} \\
W_3 & 20 \\
W_4 & 25 \\
W_5 & 30 \\
W_6 & 10 \leftarrow \text{ERROR POINTER} \\
P & 130 \\
Q & 440
\end{array}$$

SYNDROMES

$$S_1 = 10 + 20 + 20 + 25 + 30 + 10 - 130 = -15$$

$$S_2 = 60 + 100 + 80 + 75 + 60 + 10 - 440 = -55$$

$$W_2' = W_2 + E_2$$
$$W_6' = W_6 + E_6$$

$$S_1 = E_2 + E_6 = -15$$
$$S_2 = 5E_2 + E_6 = -55$$

$$\begin{array}{c|c}
E_2 = S_1 - E_6 & E_6 = S_1 - E_2 \\
5E_2 = S_2 - E_6 & E_6 = S_2 - 5E_2
\end{array}$$

$$E_2 = \frac{1}{4}S_2 - \frac{1}{4}S_1 = -10$$

$$E_6 = \frac{5}{4}S_1 - \frac{1}{4}S_2 = -5$$

CORRECTION:

$$W_2 = W_2' - E_2 = 20 - (-10) = 30$$
$$W_6 = W_6' - E_6 = 10 - (-5) = 15$$

(C) Block correction code with double parity showing double pointer correction.

Fig. 6-25. Examples of

RECEIVED DATA AND TWO PARITY WORDS

W_1 10
W_2 30
W_3 20
W_4 25 $S_1 = -20$
W_5 10
W_6 15 $S_2 = -40$
P 130
Q 440

ALGEBRAICALLY WE SEE THAT

IF $6S_1 = S_2$ THEN W_1 IS ERRONEOUS
IF $5S_1 = S_2$ THEN W_2 IS ERRONEOUS
IF $4S_1 = S_2$ THEN W_3 IS ERRONEOUS
IF $3S_1 = S_2$ THEN W_4 IS ERRONEOUS
IF $2S_1 = S_2$ THEN W_5 IS ERRONEOUS
IF $S_1 = S_2$ THEN W_6 IS ERRONEOUS
IF $S_1 \neq 0$ AND $S_2 = 0$ THEN P IS ERRONEOUS
IF $S_1 = 0$ AND $S_2 \neq 0$ THEN Q IS ERRONEOUS

IN THIS CASE $2S_1 = S_2$, W_5 IS ERRONEOUS THUS (AS IN SINGLE ERASURE CASE):

$S_1 = 10 + 30 + 20 + 25 + 0 + 15 - 130 = -30$

$W_5 = W_5' - S$
$\quad = 0 - (-30)$
$\quad = 30 \qquad$ CORRECTED

(B) Block correction code with double parity showing single error without pointer.

RECEIVED DATA AND PARITY WORDS

W_1 10

W_2 30

W_3 —

W_4 —

W_5 30

W_6 15

P 130

Q 440

SYNDROMES: $S_1 = 10 + 30 + 0 + 0 + 30 + 15 - 130 = -45$

$\qquad\qquad\quad S_2 = 60 + 150 + 0 + 0 + 60 + 15 - 440 = -155$

$W_3' = W_3 + E_3$
$W_4' = W_4 + E_4$

$S_1 = E_3 + E_4 = -45$
$S_2 = 4E_3 + 3E_4 = -155$

$E_3 = S_1 - E_4 \quad / \quad E_4 = S_1 - E_3$
$4E_3 = S_2 - 3E_4 \quad / \quad 3E_4 = S_2 - 4E_3$

$E_3 = S_2 - 3S_1 = -20$
$E_4 = 4S_1 - S_2 = -25$

CORRECTION

$W_3 = W_3' - E_3 = 0 - (-20) = 20$
$W_4 = W_4' - E_4 = 0 - (-25) = 25$

(D) Block correction code with double parity showing double erasure correction.

double parity block coding.

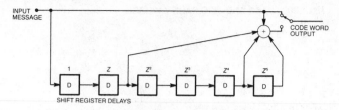

Fig. 6-26. A convolutional code encoder with six delay blocks.

6.3-4 Interleaving • Error correction depends on an algorithm's ability to efficiently use redundant data to reconstruct incorrect or lost data; when the error is sustained, as in the case of a burst error, both the data and the redundant data is lost, and correction becomes difficult or impossible. To overcome this, data words are dispersed throughout the data stream before the data is recorded. If a burst error occurs, it will damage a section of recorded data; however, upon playback the bit stream will be de-interleaved thus the data will return to its original sequence, and the errors will be distributed through the bit stream. With valid data now surrounding the damaged data, it is easier for the algorithm to reconstruct the damaged data, using information from the valid data, as shown in Fig. 6-30.

Simple delay interleaving effectively disperses data. Many block checksums work properly only if there is only one word error per block; a burst error would violate this rule; however, interleaved and de-interleaved data may very well result in one erroneous word in a given block. Thus, interleaving greatly increases burst error correctability of block codes. A single erasure code and delay interleave system suggested by Doi is shown in Fig. 6-31. Data words are dispersed to every sixth word in the data stream, thus a burst error of up to six words long could be corrected or concealed, having been turned into six random errors, given proper pointers.

Bit interleaving accomplishes much the same purpose as block interleaving; it permits long burst errors to be handled as short burst errors or random errors. Any interleaving scheme requires a buffer long enough to hold the distributed data during both interleaving and de-interleaving.

6.3-5 Cross-Interleave Code • A cross-interleave code is comprised of two or more block codes assembled with a convolutional structure. It is classified as a

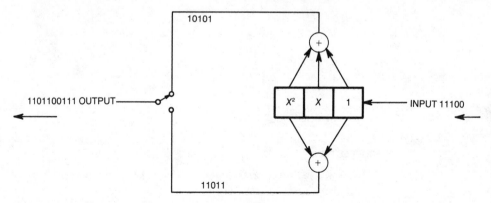

(A) Convolutional encoder with k = 3 and r = 1/2.

Fig. 6-27. Convolutional

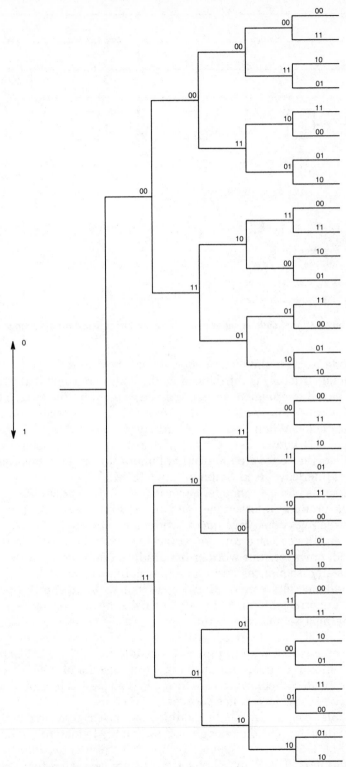

(B) Convolutional code tree diagram.

encoding.

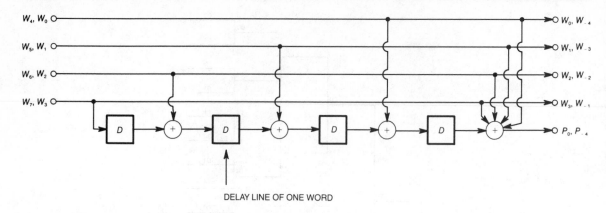

$$P_0 = W_0 + W_1 + W_2 + W_3 + W_{-4} + W_{-7} + W_{-10} + W_{-13}$$
$$P_4 = W_4 + W_5 + W_6 + W_7 + W_0 + W_{-3} + W_{-6} + W_{-9}$$
$$P_8 = W_8 + W_9 + W_{10} + W_{11} + W_4 + W_1 + W_{-2} + W_{-5}$$
$$P_{12} = W_{12} + W_{13} + W_{14} + W_{15} + W_8 + W_5 + W_2 + W_{-1}$$
$$P_{16} = W_{16} + W_{17} + W_{18} + W_{19} + W_{12} + W_9 + W_6 + W_3$$

Fig. 6-28. **Convolutional code encoder generating one check word for every four data words.**

product code because the block codes are arranged in rows and columns, the convolutional structure is introduced as the codes are separated by delay and then interleaved to enhance correctability, as shown in Fig. 6-32. The method is efficient because the syndromes from one code can be used to point to errors in the other code. When both codes are single erasure correcting codes, the resulting code is known as a Cross-Interleave Code (CIC). In the case of the Compact Disc, the codes are the Reed-Solomon code and the method is known as the Cross-Interleave Reed-Solomon Code (CIRC).

An example of a CIC encoder suggested by Doi is shown in Fig. 6-33. The delay units produce interleaving, and the modulo 2 adders provide single erasure correcting codes. Two check words are generated (P and Q) and with two single erasure codes, errors are corrected efficiently. Triple word errors can be corrected; however, four word errors produce double word errors in all four of the generated sequences and correction is impossible. The CIC enjoys the high performance of a convolutional code, but without error propagation because any uncorrectable error in one sequence always becomes a one word error in the next sequence and thus can be easily corrected. A general block diagram for a CIC encoder/decoder is shown in Fig. 6-34. The performance of CIC can be augmented with Improved Cross-Interleave Code (ICIC) methods in which the amount of delay interleaving is increased and made irregular, and more sophisticated check words are formed, as suggested by Doi and shown in Fig. 6-35. The DASH format has adopted a CIC code.

Cross-interleave codes can be modified to service recording methods using rotary head designs. The encoder chosen for the EIAJ video rotary head format, shown in Fig. 6-36, uses a double erasure correction code, interleaving, and CRCC for pointers. The P and Q check words are generated for every six data words; P is the modulo 2 addition of all data words and Q is a weighted summation. Following interleaving, CRCC is encoded for the resultant eight

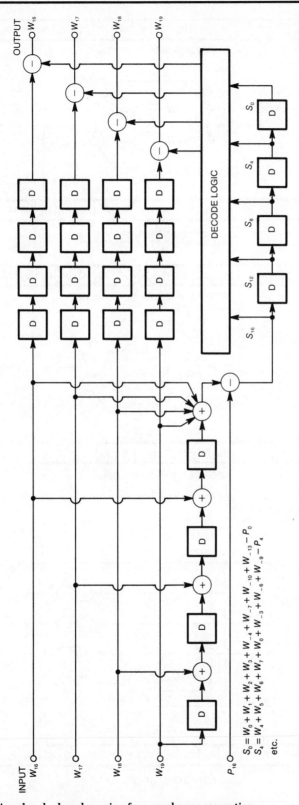

Fig. 6-29. Convolutional code decoder using four syndrome correction.

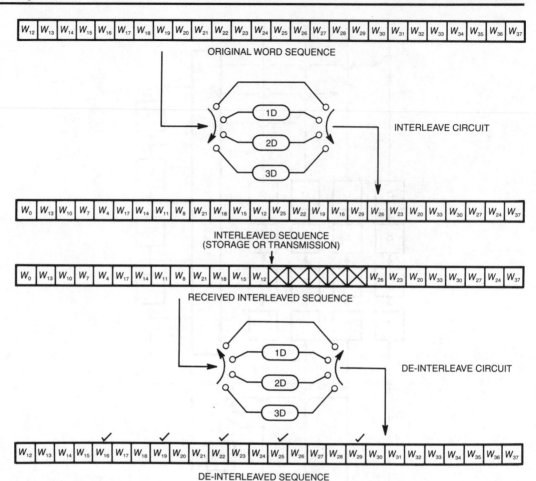

Fig. 6-30. Zero, one, two, and three word delays perform interleaving and de-interleaving for error dispersion/correction.

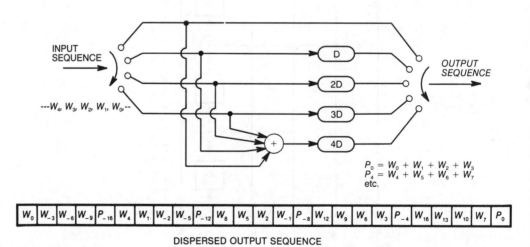

$$P_0 = W_0 + W_1 + W_2 + W_3$$
$$P_4 = W_4 + W_5 + W_6 + W_7$$
etc.

Fig. 6-31. Single erasure interleaved encoder.

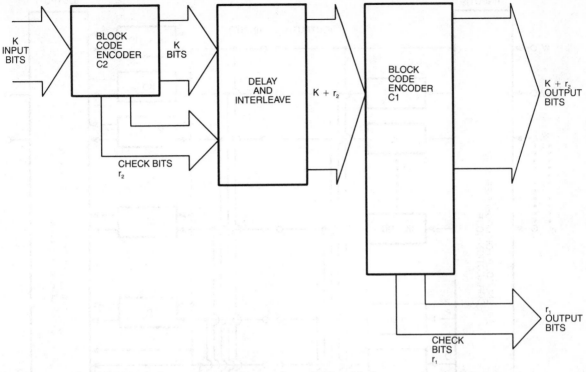

Fig. 6-32. Cross-interleave code encoder. Syndromes from the first block are used as error pointers in the second block.

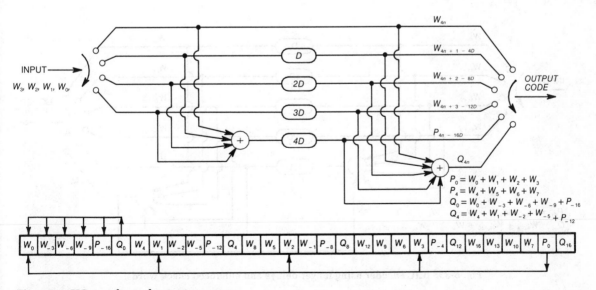

$$P_0 = W_0 + W_1 + W_2 + W_3$$
$$P_4 = W_4 + W_5 + W_6 + W_7$$
$$Q_0 = W_0 + W_{-3} + W_{-6} + W_{-9} + P_{-16}$$
$$Q_4 = W_4 + W_1 + W_{-2} + W_{-5} + P_{-12}$$

Fig. 6-33. CIC encoder and output sequence.

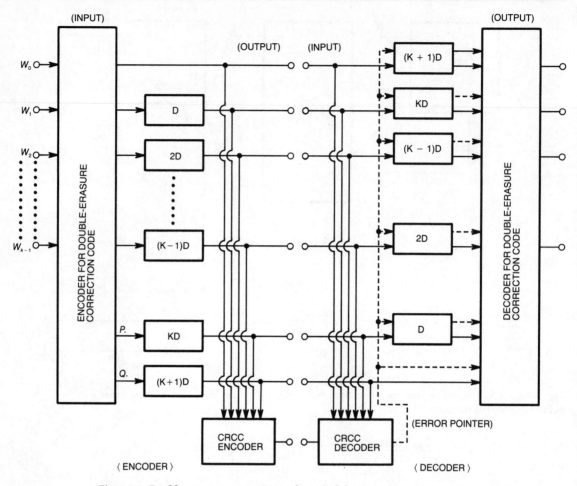

Fig. 6-34. Double erasure correction code with delay interleave and CRCC.

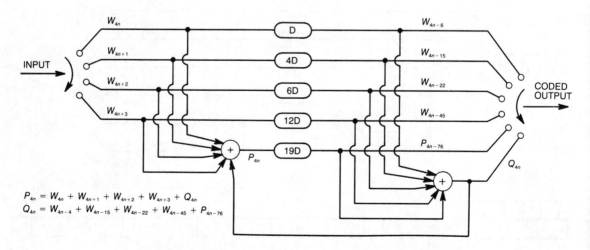

$$P_{4n} = W_{4n} + W_{4n+1} + W_{4n+2} + W_{4n+3} + Q_{4n}$$
$$Q_{4n} = W_{4n-4} + W_{4n-15} + W_{4n-22} + W_{4n-45} + P_{4n-76}$$

Fig. 6-35. ICIC encoder using longer delays and enhanced check words.

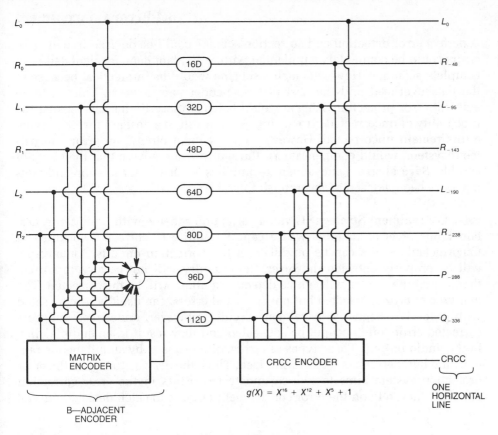

$$g(X) = X^{16} + X^{12} + X^5 + 1$$

Fig. 6-36. EIAJ encoder used in rotary head recorders.

words; this forms data blocks for each horizontal line. The double erasure code can correct two words per block prior to delay; the recorded code can be corrected for burst errors up to 32 blocks long.

Many types of error correction exist for many types of applications; designers must judge correctability of random and burst errors, redundancy overhead, probability of misdetection, and maximum burst error lengths to be corrected, or concealed. For digital audio, errors which cannot be corrected are concealed; however, a misdetected error often cannot be concealed and this could result in a click in the musical output of the system. Error correction system designers must ensure that this does not happen under normal worst-case conditions. Other design criteria are set by the particular application; for example, open reel digital tape recorders are susceptible to burst errors from fingerprints whereas the Compact Disc is relatively tolerant to fingerprints because the transparent layer places them out of focus to the optical pickup. Because of the differing nature of errors, some predictable, and some, such as those produced by misaligned recorders and players, unpredictable, error correction systems ultimately must employ differing techniques to guard against them. Delay interleaving, CRCC, and CIC are all employed to successfully correct most types of errors found in digital audio recordings; performance, redundancy, and cost of encoder and decoder must be balanced for efficient error correction.

6.4 Error Concealment

A perfect error detection and correction scheme could be devised in which all errors could be completely supplanted with redundant data or calculated with complete accuracy. However, such a scheme would be impractical because of the data overhead and the cost of the encoder and decoder. Thus, a more efficient error protection scheme would balance those limitations against the probability of uncorrected errors, and would result in a design in which severe errors remain uncorrected. However, a subsequent circuit, an error concealment system, would compensate for those errors and ensure that they are not audible. Several error concealment techniques, such as interpolation and muting, have been devised to accomplish this.

6.4-1 Concealment Strategy • Given a correction scheme with limitations, certain kinds of error patterns will escape its ability to correct. Two kinds of uncorrectable errors can be output from the correction circuits: some errors will be properly detected; however, the circuitry will not be able to correct them, and other errors will not be detected at all, or will be miscorrected. The first type of errors, detected but not corrected errors, can usually be concealed with properly designed concealment circuits. However, undetected and miscorrected errors often cannot be concealed and may result in an audible click in the audio output. These types of errors, often caused by simultaneous random and burst errors, must be minimized. Thus, the design strategy of the error protection system aims to reduce undetected errors in the error correction circuitry, then rely on the error concealment circuits to catch detected but not corrected errors.

6.4-2 Interpolation • Following de-interleaving, most errors, even burst errors, are interspersed with valid data words. It is thus reasonable to employ techniques in which surrounding valid data is used to calculate new data to replace the missing or uncorrected data. This technique works well provided that errors are sufficiently dispersed, and there is continuity to the data values. Fortunately, digital data comprising a musical selection can often undergo interpolation without becoming audible. Interpolation techniques perform concealment with great success in digital audio recording because of the correlation between sample points.

In its simplest form, interpolation simply holds the previous data value, and repeats it to cover the missing or incorrect word. This is called zero-order or previous-value interpolation, and is shown in Fig. 6-37A. In first-order interpolation, sometimes called linear order interpolation, the erroneous word is replaced with a newly calculated word derived from the mean value of the previous and subsequent word, as shown in Fig. 6-37B. In many digital audio systems, a combination of zero and first-order interpolation is used. If consecutive word errors occur in spite of interleaving, then previous value interpolation is used to replace consecutive errors, but the final held word's value is calculated from the mean value of the held and subsequent word. If the errors are random, that is, valid words surround the errors, then mean value calculations are used. Although it is nonintuitive, studies have shown that the length of an error does not overly affect perception of the error as long as the interpolated words are surrounded by valid data. Other higher order interpolation is

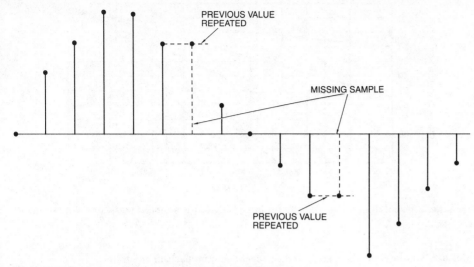

(A) Zero order interpolation—previous value held.

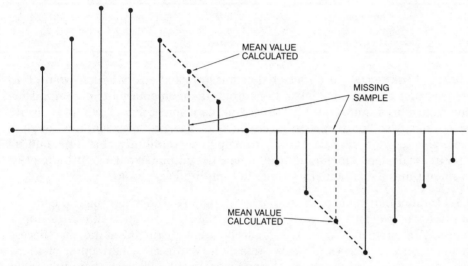

(B) First order interpolation—mean value calculated.

Fig. 6-37. Interpolation method of error concealment.

sometimes used, nth-order interpolation uses a higher-order polynomial to calculate substituted data; third and fifth order interpolation are sometimes used. The relative values of interpolation noise as demonstrated by Doi, are shown in Fig. 6-38.

Interpolation calculations can become quite sophisticated; besides a penalty of increased hardware costs, the processing must be accomplished quickly enough to maintain the data rate. A summary of burst error correction and concealment statistics for three formats is presented in Table 6-2.

6.4-3 Muting • Muting is the simple process of setting the value of the missing or uncorrected words to zero. This silence is preferable to the unpredictable sounds which could result from decoding incorrect data. Muting might be

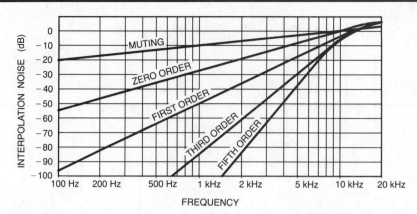

Fig. 6-38. **Interpolation noise (for 100 bit burst length, pure tone signals, and uncorrectable errors).**

Table 6-2. **Burst Error Correction and Concealment Comparison of Formats**

Format	Type	Correction	Good Concealment	Marginal Concealment
EIAJ	rotary head tape	4096 bits	—	8192 bits
DASH	stationary head tape	8640 bits	33,982 bits	83,232 bits
CD	optical disc	3874 bits	13,282 bits	15,495 bits

employed in the case of undetected or miscorrected errors which would otherwise cause an audible click at the output; the momentary increase in distortion from a brief mute could be imperceptible whereas a click would typically be audible. Also, in the case of severe data damage, or player malfunction, it is preferable to mute the data output. To help avoid audibility of a mute, muting circuitry is designed to gradually attenuate the output signal's amplitude prior to a mute, and then gradually restore the amplitude afterward.

6.4-4 Duplication • One of the promises of digital audio recording is the opportunity to copy recordings without the inevitable degradation of analog duplication. Although duplication is much better with digital audio, its success depends on the success of error protection. While error correction methods provide completely correct data, error concealment does not; even under good circumstances, error concealment techniques do introduce subtle differences into copied data. Thus, subsequent generations of digital copies could contain an accumulation of concealed errors not present in the original. As a result, errors must be countered at their origin. Thus, routine precautions of clean media, clean machine mechanisms, and properly aligned circuitry are important with digital audio duplication, particularly when many generations of copying are anticipated. Definitive studies of the audibility of concealed errors, and their propagation through duplication, have not yet been accomplished.

Chapter 7

The Compact Disc

Introduction

The Compact Disc system is perhaps the most remarkable development in audio reproduction technology since the technology's birth in 1877 with Edison's invention of the wax cylinder. It embodies many revolutionary steps in design, such as digital signal storage, optical scanning, error correction, and new manufacturing processes for both players and discs; altogether it establishes a new fidelity standard for the consumer. As remarkable as the Compact Disc system is, it is only the beginning of the digital era in consumer digital audio technology.

7.1 The Compact Disc Medium

The Compact Disc system contains several unique aspects original to the audio field; when combined they form an unprecedented means of sound reproduction. A Compact Disc contains digitally encoded music which is read by a laser beam, as shown in Fig. 7-1; since the laser is focused on a reflective layer

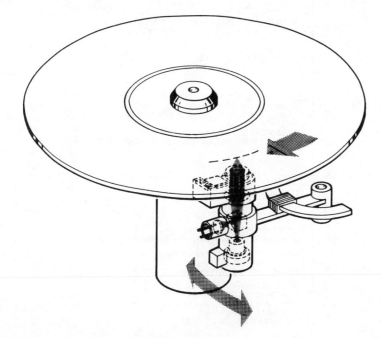

Fig. 7-1. **Compact Disc and laser pickup.**

embedded within the disc's substrate, dust and fingerprints on the reading surface will not overly affect reproduction. The effect of most errors which could normally occur can be minimized by error correction circuitry. Since no stylus touches the disc surface, there is no disc wear, no matter how often the disc is played. Thus, digital storage, error protection, and disc longevity result in a robust high fidelity audio medium.

7.1-1 History of the Compact Disc • Whereas Edison's wax cylinder was invented and marketed with seeming ease, the chronology of events in the development of the Compact Disc spans almost a decade from inception to introduction, as shown in Fig. 7-2. The Compact Disc incorporates many technologies pioneered by many individuals and corporations; however, Philips Corporation of the Netherlands and Sony Corporation of Japan must be credited with its primary development. Optical disc technology developed by Philips and error correction techniques developed by Sony, when merged, resulted in the successful Compact Disc format. The standard established by those two companies guarantees that discs and players made by different manufacturers are absolutely compatible.

Philips first studied the possibility of storing audio material on an optical disc in 1974; analog modulation methods, used for video storage were deemed unsuitable, and the possibility of digital signal encoding was examined. Furthermore, Philips established small disc diameter as a design prerequisite.

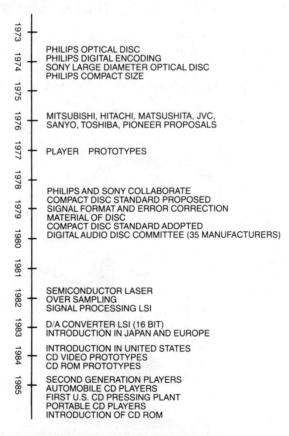

Fig. 7-2. **Chronology of the development of the Compact Disc.**

Sony similarly had explored the possibility of an optical, large diameter audio disc, and had extensively researched the error processing requirements for a practical realization of the system. Other manufacturers, such as Mitsubishi, Hitachi, Matsushita, JVC, Sanyo, Toshiba, and Pioneer had advanced proposals for a digital audio disc. In 1979 Sony and Philips reached an agreement in principle to collaborate, and with decisions on signal format and disc material, jointly proposed the Compact Disc Digital Audio system in June of 1980 which was adopted by the Digital Audio Disc Committee, a group representing over 35 manufacturers.

Following development of a semiconductor laser pickup and Large Scale Integration integrated circuits for signal processing and D/A conversion, the Compact Disc system was introduced in October of 1982 in Japan and Europe. In March of 1983, the Compact Disc was made available in the United States. Initial player prices of $900 dropped to $400 by October with over 20 brands of players being available. Over 350,000 players and 5.5 million discs were sold worldwide in 1983, and 900,000 players and 17 million discs in 1984, making the CD the most successful electronic product ever introduced. Automobile and portable players have been added to product lines, and players selling for less than $200 are available.

7.1-2 Compact Disc System Overview • The Compact Disc system is a highly efficient information storage system. Each disc stores a stereo audio signal comprised of two 16 bit data words sampled at 44.1 kHz, thus 1.41 million bits per second of audio data are output from the player. However, other overhead processing such as error correction, synchronization, and modulation are required which triple the number of bits stored on a disc. Altogether, the channel bit rate, the rate at which data is read from the disc, is 4.3218 million bits per second. A disc containing an hour of music thus holds about 15.5 billion channel bits of information, of which 5 billion bits are audio data; that is quite a feat for a disc less than 4¾ inches across. Maximum disc playing time (strictly according to legend) was determined after Philips consulted German conductor Herbert von Karajan; he advised them that a disc should be able to hold his performance of the Beethoven Ninth Symphony without interruption. A Compact Disc has a maximum playing time of 74 minutes, 33 seconds.

Information is contained in pits impressed into the disc's plastic substrate, that surface is then metalized to reflect the laser beam used to read the data from underneath the disc. A pit is about 0.5 micrometer wide, and a disc might hold about 2 billion of them; if a disc were enlarged such that its pits were the size of grains of rice, the disc would be a half mile in diameter. In fact, the pits are among the smallest manufactured formation. The pits are aligned in a spiral track similar to the spiral groove in a conventional record; however, the CD pit track runs from the inside diameter of the disc to the outside. If un-spiraled, a CD track would stretch about 3 miles. The pitch of the CD spiral, that is, how close together successive turns are, is much greater than on a conventional record; 30 CD pit laps would fit in the width of a human hair, or approximately 50 micrometers. Of course, the CD also differs from the LP in that the information is digitally encoded; each pit edge represents a binary 1, flat areas on land between pits or areas inside pits are decoded as binary 0. Data is read from the disc as a change in intensity of reflected laser light.

Compact Disc players parallel conventional LP turntables, but the similarity is limited. The disc is rotated by a motor, but since the pit spiral is recorded as a constant velocity (1.2 to 1.4 meters per second depending on program length), the player must adjust its own rotational speed to maintain a constant velocity as the spiral diameter changes, thus speed of rotation of a disc varies from about 8 revolutions per second to about 3.5 revolutions per second.

An optical pickup replaces a phonograph's mechanical stylus; a laser beam is emitted, and is guided through optics to the disc surface. The reflected light is detected by the pickup, and the data from the disc carried on that beam is converted to a digital electrical signal. Since nothing touches the disc, except light, light itself, and electrical servo circuits, are used to keep the laser beam properly focused on the disc surface, and properly aligned with the spiral track. The pits are encoded with Eight-to-Fourteen Modulation (EFM) for greater storage density, and Cross-Interleave Reed-Solomon Code (CIRC) for error correction; circuits in players provide demodulation and error correction. Because of the complexity of the processing, most players contain several microprocessor chips, or special integrated circuits containing microprocessor elements. When the actual audio data has been properly recovered from the disc and converted into a binary signal, it is input to D/A converters for conversion into an analog signal, then passes through output filters, and output amplifiers to provide the final analog audio signal compatible with home stereo systems. A functional block diagram of a player showing disc, motor, optical reader, servo for auto-tracking and auto-focusing control, and audio signal decoding is shown in Fig. 7-3.

The Compact Disc system delivers high fidelity sound, with outstanding performance specifications. With 16 bit quantization, sampled at 44.1 kHz, players typically exhibit a frequency response of 5 Hz to 20 kHz with a deviation of ± 0.5 dB. Dynamic range is greater than 90 dB, signal-to-noise ratio is greater than 90 dB, and channel separation is greater than 90 dB at 1 kHz. Harmonic distortion is less than 0.004 percent at 1 kHz. Wow and flutter is limited to the tolerances of quartz accuracy, which is essentially unmeasurable.

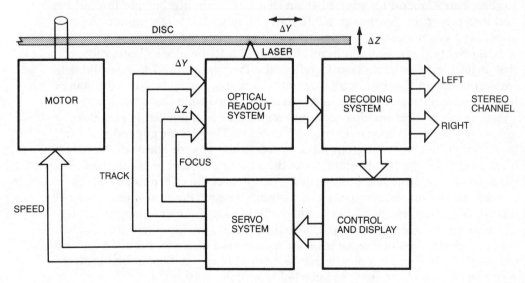

Fig. 7-3. **Functional block diagram of CD player.**

With digital filtering, phase shifts are less than 0.5 degree angle of deviation. Excluding unreasonable abuse, a disc will remain in perfect playing condition indefinitely since there is no pickup touching the disc, or aging of the medium. A typical system specification sheet is presented in Fig. 7-4.

7.2 The Disc

A Compact Disc is a very sophisticated form of data storage. A typical disc can hold 5 billion bits of audio data and thus in storage capacity a CD is far superior to magnetic floppy disks. The longevity of a Compact Disc makes it preferable to other media prone to higher error rates or total failure as aging occurs. For the consumer, the Compact Disc is a highly convenient medium, featuring both small size and immediate access to any part of the stored program. The Compact Disc is conducive to mass production. Unlike tapes which must be cycled through their entire length during duplication, Compact Discs may be formed wholly.

System	Compact Disc Digital Audio
Audio Performance (20 Hz-20 kHz unless otherwise stated)	
Frequency range	20 Hz-20 kHz, ± 0.3 dB
Phase linearity	±0.5°
Signal to noise ratio (dynamic range)	greater than 90 dB
Channel separation	greater than 86 dB greater than 90 dB (at 1 kHz)
T.H.D. (incl. noise)	less than 0.005% less than 0.004% (at 1 kHz)
Intermodulation distortion measured at max. output level	less than -86 dB
Out-band rejection (frequencies more than 24 kHz)	greater than 50 dB
Wow and flutter	Quartz crystal precision
Optical read-out system	
Laser type	Semiconductor Al Ga As
Numerical aperture	0.45
Wave length	790 nm
Output	
Max. output level (at MSB)	2 V rms, typical
Output impedance	less than 100 Ohms
Minimum load impedance	10 kOhms
Power Supply	
Supply voltages	120 V AC
Supply frequency	60 Hz
Power consumption	30 W approx.

Fig. 7-4. **Typical performance specifications of a CD player.**

7.2-1 Disc Specifications • The physical characteristics of a Compact Disc are summarized in Fig. 7-5. It has an outer diameter of 120 millimeters, a center hole 15 millimeters across, a thickness of 1.2 millimeters, and a weight of about 14 grams. Each data pit is about 0.5 micrometer wide and its length, and the length of the land separating each pit, varies incrementally from 0.833 to 3.054 micrometers. It is that varying ratio of pit and land which encodes the data itself. Other specifications define track pitch, the distance between successive tracks, to be 1.6 micrometers, with maximum eccentricity of track radius to be 70 micrometers, and a maximum angular deviation (skew) of 0.6 degree.

A Compact Disc is a single-sided medium; manufacturers could make double-sided discs, but the cost would be more than the cost of two single-sided discs. Thus one side merely holds the printed label, while the data pits are viewed through the opposite side of the substrate. Because of disc construction, the laser beam must pass through the disc's transparent thickness to reach the pits. The beam, output from a semiconductor device, strikes the underside of the disc (where the underside of the pits appear as bumps). As it travels through the disc's thickness, the 1.5 refractive index of the substrate causes the beam diameter to focus from its 0.8 millimeter width at the substrate surface to a focal spot of 1.7 micrometers at the reflective pit surface. This focusing causes any substrate imperfections, such as scratches or fingerprints, to appear to be out of focus; their shadow is a small fraction of its original size when it reaches the disc surface thus its effect is minimized. As the laser beam strikes pits and the reflective areas between, the returning variable intensity beam carries the information on the disc back to a photodiode and then to an EFM modulator where the original binary signal is restored for D/A conversion and output as an analog voltage.

Although the pits and the intervening reflective surface on the disc surface hold the binary data, they do not directly designate 1s or 0s. Rather, each pit edge whether leading or trailing is a 1 and all increments in between, whether inside or outside a pit are 0s. This is a much more efficient storage technique than coding the binary digits directly with pits. The technique is illustrated in Fig. 7-6. For example, 4 pits are being used to code 31 bits of information thus saving valuable disc space and permitting longer playing time. In addition, EFM facilitates the recovery of a clock signal used to control the disc motor speed. But even this information, known as channel bits, does not directly represent audio data, rather, data frames are formed which carry audio data, as well as error correction and other information. These frames are assembled during the encoding process.

7.2-2 Data Encoding • The channel bits, the data actually encoded on the disc, is the end product of an elaborate coding scheme accomplished prior to disc mastering then decoded each time the disc is played. The block diagram for CD encoding is shown in Fig. 7-7. The data begins, of course, as music. Whether it is an analog recording, delta modulation, or pulse code modulation, it is converted to PCM, the familiar system in which the signal is time sampled and amplitude quantized into a parallel binary number, as reviewed in Fig. 7-8. The PCM format is recorded using a professional digital audio processor and a video cassette recorder. The master tape contains the following information: Video format tracks with digital audio data, analog audio channel 1 with PQ subcode, and analog audio channel 2 with continuous SMPTE (non-droop

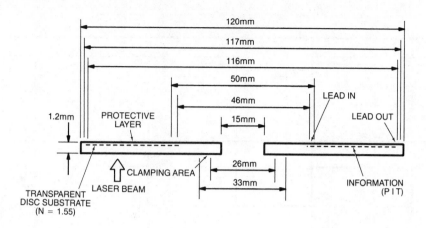

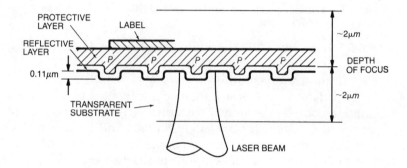

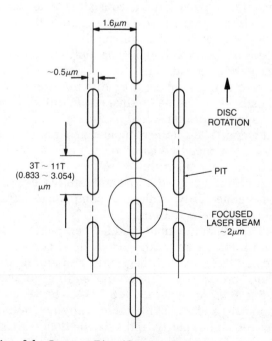

Fig. 7-5. **Physical characteristics of the Compact Disc. (Courtesy Sony)**

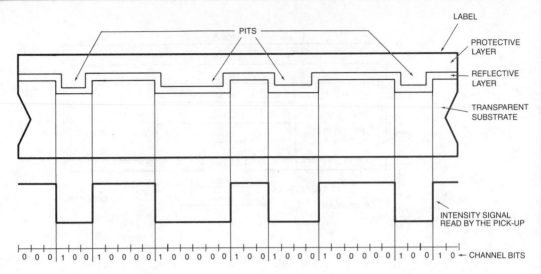

LABEL

PROTECTIVE LAYER

REFLECTIVE LAYER

TRANSPARENT SUBSTRATE

INTENSITY SIGNAL READ BY THE PICK-UP

0 0 0 | 1 0 0 | 1 0 0 0 0 | 1 0 0 0 0 0 | 1 0 0 | 1 0 0 0 | 1 0 0 0 0 0 | 1 0 0 | 1 0 ← CHANNEL BITS

Fig. 7-6. **Transitions between pit bottom and disc surface, or "land" represents a binary 1, flats represent binary 0s.**

frame) time code. Once the audio data is in the digital domain, it must undergo CIRC error correction encoding and EFM modulation, and subcode and synchronization words must be incorporated as well.

All data on a CD is formatted by frames; by definition a frame is the smallest complete section of recognizable data on a disc. The frame provides a means to distinguish between audio data and its parity, the synchronization word, and the subcode; frame construction prior to EFM modulation is shown in Fig. 7-9. During encoding, prior to CD mastering, all of the previous data is placed into the frame format. The end result of encoding and modulation is a bit stream of frames, each frame consisting of 588 channel bits. It is the channel bits themselves which are physically cut into the disc.

To begin assembly of a frame, six 32 bit PCM audio sampling periods (alternating from left and right channel) are grouped in a frame. The 32 bit sampling periods are divided to yield four 8 bit audio symbols. To scatter possible errors, the symbols from different frames are interleaved so that the audio signals in one frame originate from different frames. In addition, eight 8 bit parity symbols are generated per frame, four in the middle of the frame, and four at the end. The interleaving and generation of parity bits constitutes the error correction encoding based on the Cross-Interleave Reed-Solomon Code. The CIRC encoding algorithm is shown in Fig. 7-10. It can be seen that bits from the audio signal are delayed and interleaved, and two encoding stages, C1 and C2, generate parity symbols. A complete examination of CIRC decoding is presented in the next section.

One PQ subcode symbol is added per frame; these subcode user bits (P and Q) contain information for the disc table of contents detailing total number of selections on the disc, their beginning and ending points, index points within a selection, program beginning and ending points, and preemphasis on/off control. Six of these eight bits (R,S,T,U,V, and W) are available for other applications, such as encoding video information on audio CDs.

After the audio, parity, and subcode data is assembled, the data is modulated using EFM. This gives the bit stream specific patterns of 1s and 0s, thus

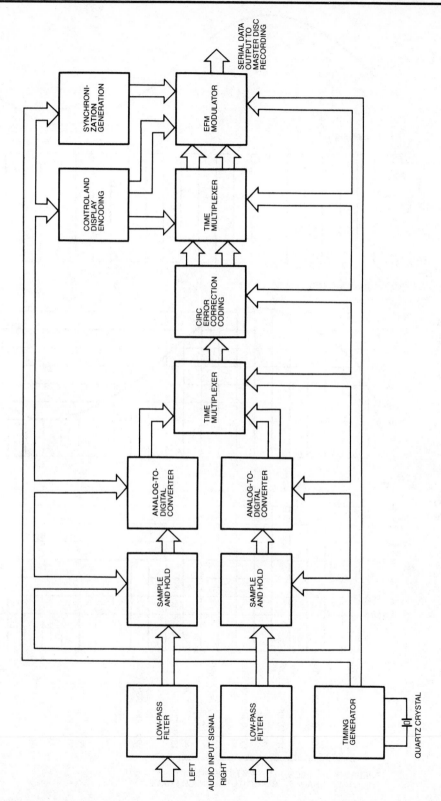

Fig. 7-7. **CD encoding system.**

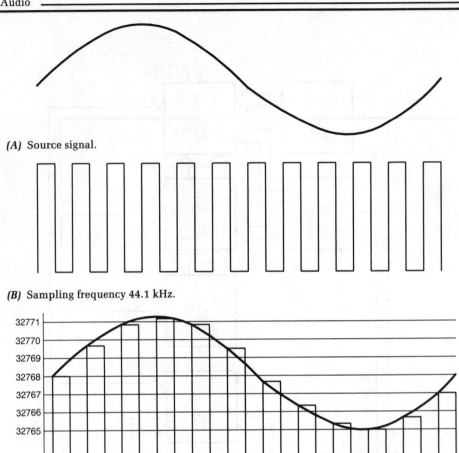

(A) Source signal.

(B) Sampling frequency 44.1 kHz.

(C) Sampling.

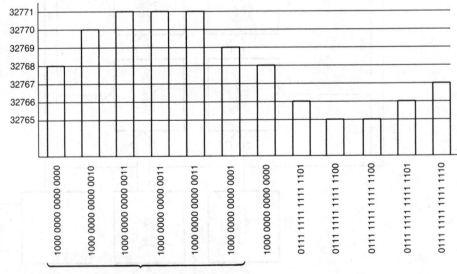

THESE SIX SAMPLES, PLUS SIX MORE FROM
THE OTHER CHANNEL, CONSTITUTE ONE CD FRAME

(D) Quantization and codification to binary 16 bits.

Fig. 7-8. PCM digitization process results in data ready for encoding prior to CD mastering.

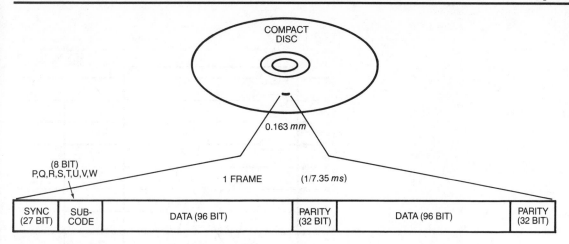

Fig. 7-9. **CD recording format showing frame format.**

defining the lengths of pits and lands to facilitate optical reading of the disc. Eight-to Fourteen Modulation results in a pit and land pattern with the maximum number of transitions possible with a minimum and maximum pit and land length; in part, this helps facilitate control of the spindle motor speed. A straight 8 bit code would result in unmanageable pattern violations. To accomplish EFM, blocks of 8 bits are translated into blocks of 14 bits using a dictionary which assigns an arbitrary and unambiguous word of 14 bits to each 8 bit word, as shown in Fig. 7-11. Blocks of 14 bits are linked by three merging bits to maintain the proper pattern between words, aid clock synchronization, and suppress any low frequency component. With the addition of merging bits, the ratio of bits before and after modulation is 8:17. The resultant channel stream produces pits and lands which are at least 2, but no more than 10 successive 0s long, as shown in Fig. 7-12. After EFM, there are thus more bits to accommodate, but acceptable land and pit patterns become available. Specifically, the 8 bit symbols require $2^8 = 256$ unique patterns, and of the possible $2^{14} = 16,384$ patterns in the 14 bit system, 267 meet the pattern requirements, thus 256 are utilized and 11 discarded. With this modulation the highest frequency in the signal is decreased, therefore, a lower track velocity can be utilized. One benefit is conservation of disc real estate. With EFM, a 25 percent gain in data density is achieved over unmodulated coding.

The resulting EFM data must be delineated thus a synchronization pattern is placed at the beginning of each frame. The synchronization word is uniquely identifiable from any other possible data configuration (specifically the 24 bit synchronization word is 100000000001000000000010 plus three merging bits). With the synchronization pattern, the pickup can always identify data frames. The frame assembly thus contains a 27 bit sync word, 17 bits of subcode, 12 words of 17 bit audio data, 4 words of 17 bit parity, 12 more audio words, and 4 more parity words for a total of 588 bits per frame. Since each 588 bit frame contains twelve 16 bit audio samples, the result is 49 channel bits per audio sample. Thus when the data manipulation is completed, the original audio bit rate of 1.41 million bits per second has been augmented to 4.3218 million channel bits per second, such is the price of overhead. That resulting bit stream is encoded onto the disc as channel bits. The entire encoding process is shown in Fig. 7-13.

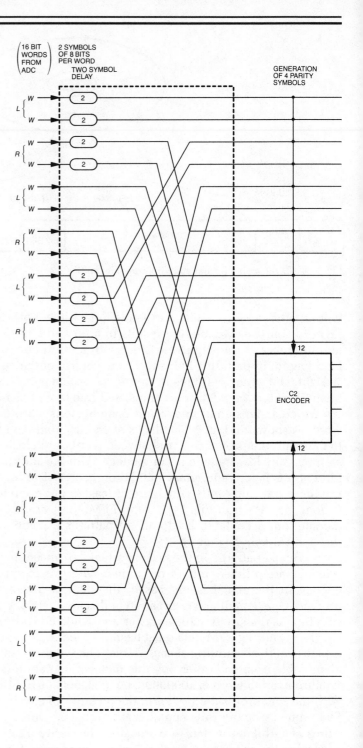

Fig. 7-10. CIRC

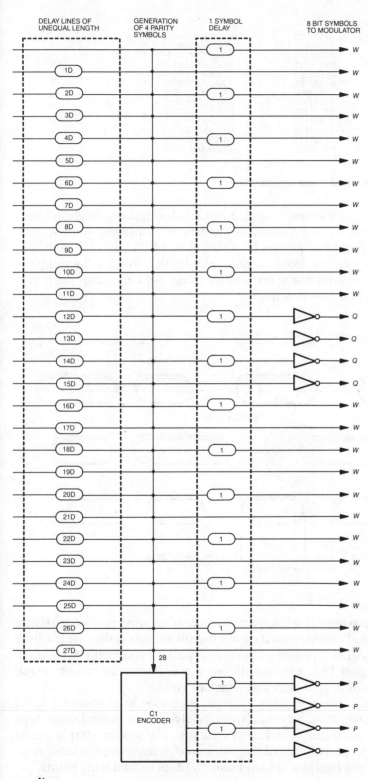

encoding.

DATA	EFM
01100100	01000100100010
01100101	00000000100010
01100110	01000000100100
01100111	00100100100010
01101000	01001001000010
01101001	10000001000010
01101010	10010001000010
01101011	10001001000010
01101100	01000001000010
01101101	00000001000010
01101110	00010001000010
01101111	00100001000010
01110000	10000000100010

Fig. 7-11. **Excerpts from an EFM conversion table.**

The Compact Disc bit stream is thus more complex than one might suspect. In the interests of density and robustness the data must undergo some sophisticated processing. Eight-to-Fourteen Modulation and CIRC, as well as the entire frame structure used to delineate data, require that the data be uniquely packaged before it is encoded onto the Compact Disc. And of course, the player is left with the job of deciphering it all.

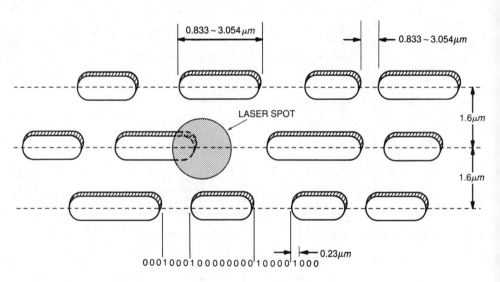

Fig. 7-12. **Pits.**

7.2-3 Disc Manufacturing • The Compact Disc manufacturing process is similar to that of conventional records thus it enjoys the advantages of the disc medium in which the information is placed on the disc simultaneously with its creation. However, the Compact Disc requires different manufacturing processes and greater quality control to guarantee a satisfactory yield.

Although manufacturers employ different processes to produce Compact Discs, the CD manufacturing process always involves three general steps: tape mastering, CD disc mastering, and disc duplication. The tape mastering can be accomplished in a properly outfitted recording studio; however, the latter steps require the specialized equipment found only in CD manufacturing plants.

The tape mastering process is the culmination of the recording process, in

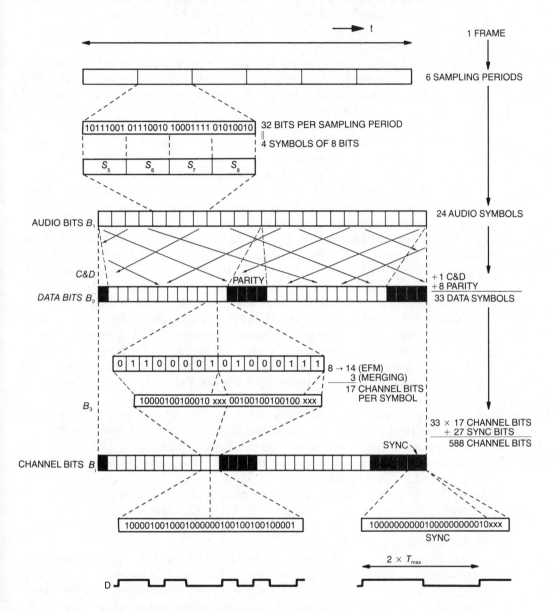

Fig. 7-13. Bit streams in the encoding system. (Courtesy Philips)

which the master audio tape (hopefully digital) has been edited, and recorded on a ¾ inch format video recorder via a professional digital audio processor. All CDs are manufactured from the video PCM format; however, it is possible to transfer from an incompatible digital recording format. Although the digital recording could be converted to analog, and then to the video PCM format, degradation would result hence a trans-coder may be employed for a digital-to-digital transfer without deterioration in signal quality. During mastering, the user bits, sometimes called the PQ subcode, is uniquely created for the master using a subcode editor. The subcode is entered on the video PCM tape and along with the music data is later modulated into the CD format during disc cutting. This preparation of the master tape can be carried out in recording studios. The remainder of the CD manufacturing chain is accomplished at the CD pressing plant.

Compact Disc mastering is the first disc manufacturing process; these steps are shown in Figure 7-14. A glass plate about 240 millimeters in diameter and 6 millimeters thick comprised of simple float glass is washed in alkali and Freon, lapped, and polished with a CeO_2 optical polisher; the plate is prepared in a clean room with extremely stringent dust filtering. After inspection and cleaning, the plate is tested for optical drop-outs with a laser; any burst drop-outs in reflected intensity are cause for rejection of the plate. An adhesive is applied, followed by a coat of photoresist applied by a spinning developer machine. The depth of the photoresist coating is critical; it ultimately determines the pit depth. The plate is cured in an oven and stored with a shelf life of several weeks. The plate is ready for master cutting.

The cutting machine is a million dollar device which cuts the data spiral into the master glass disc. The cutting machine has a control rack consisting of a minicomputer with video terminal and floppy disk drive, video transport, PCM audio processor, and diagnostic equipment. The lathe uses a 15 milliwatt, 441.6 nanometer wavelength Helium-Cadmium cutting laser which is intensity modulated by an acousto-optic modulator to create the cutting signal corresponding to the data on the audio master tape; another laser which does not affect the master disc photoresist is used for focus and tracking. The master glass plate coated with photoresist is placed on the lathe, and exposed with the cutting laser to create the spiral track, cutting the disc contents in real time as the master tape is played through the PCM processor. Quality of the production discs is directly dependent on the master cutter's signal characteristics, such as eye pattern symmetry, signal modulation amplitude, and track following. To guard against disc contamination, stringent air filtering is used inside the lathe cutting bay. Although the optics are similar to those found inside consumer CD players (polarized beam splitter, objective lens, semiconductor laser) the mechanisms are built on a grander scale, especially in terms of isolation from vibration; for example, the stylus block is supported and moved by an air-float slider. The entire cutting process is accomplished automatically, under computer control.

After exposure in the master cutter, the glass master is developed by an automatic developing machine. Developing fluid washes the rotating disc surface, etching away the exposed areas of photoresist. During development a laser monitors photoresist depth and stops development when proper engraving depth has been reached, that is, when the etching reaches the glass substrate, the pit depth thus depends on the thickness of the photoresist layer. In theory, the optimum signal from the finished CD results when the pit depth equals one-quarter the wavelength in the transparent substrate of the 790 nanometer laser used in CD players. Pit depth, one-fourth of 790 nanometers, divided by the refraction index of 1.5, is theoretically 132 nanometers. Furthermore, the pit depth dictates that the intensity of the light reflected from the pit bottom equals the intensity of the light reflected from the surface, thus destructive interference causes an absence of reflected light wherever there is a pit, distinguishing it from the almost total reflection from the land surrounding the pits. In practice, a compromise must be made to balance the need for zero reflected

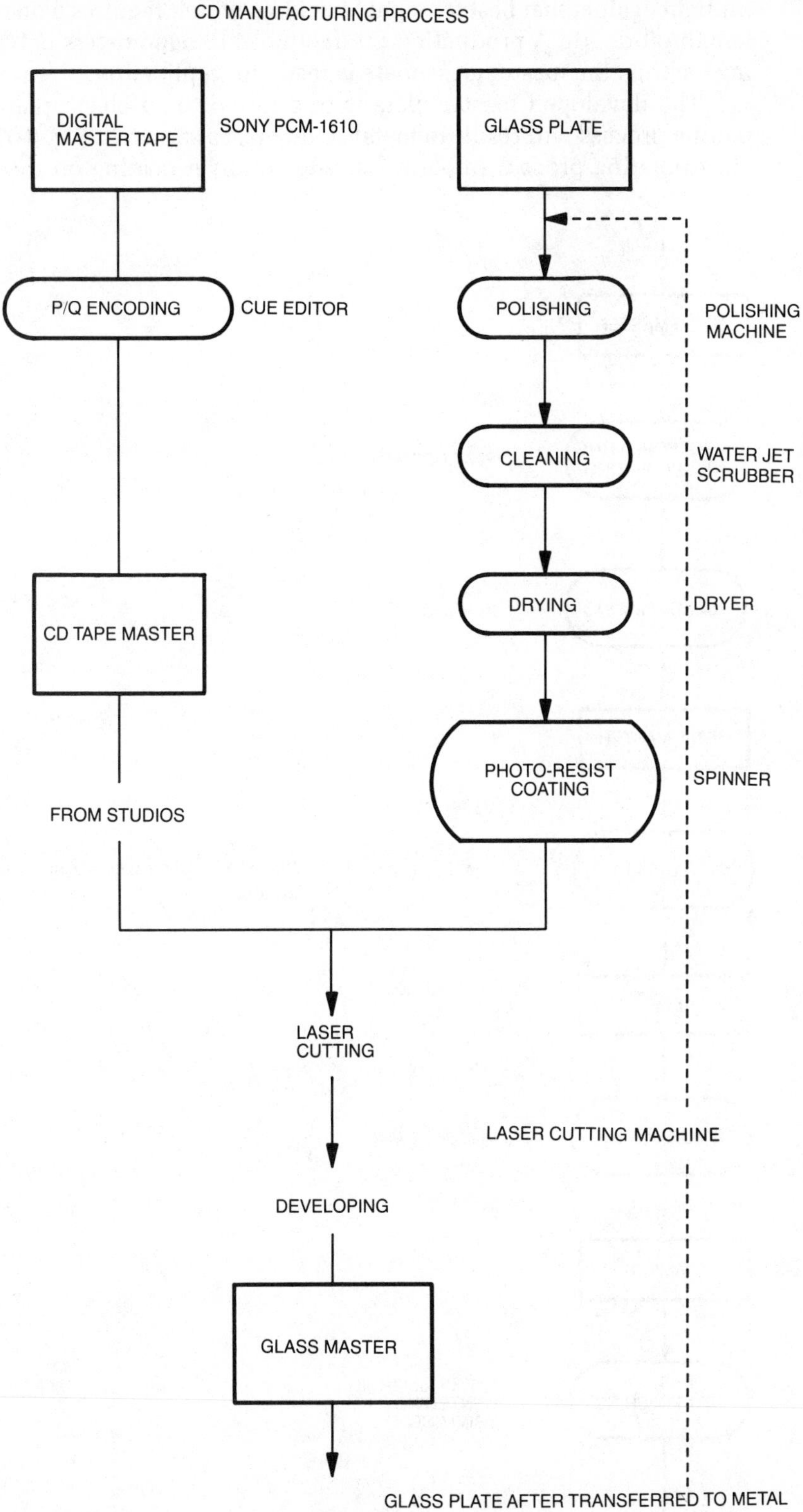

CD MANUFACTURING PROCESS

Fig. 7-14. Disc mastering. The CD master tape is used to cut a master glass disc.

pit light against that best for signal tracking, which requires a one-eighth wavelength pit depth. A production pit depth of 110 nanometers is typical. Following etching, the master glass plate is ready for replication.

The developed master plate is transferred to an electroplating room; the plating process will result in metal stampers, as shown in Fig. 7-15. The master electroplating process imparts a sputtered silver coating on the glass master.

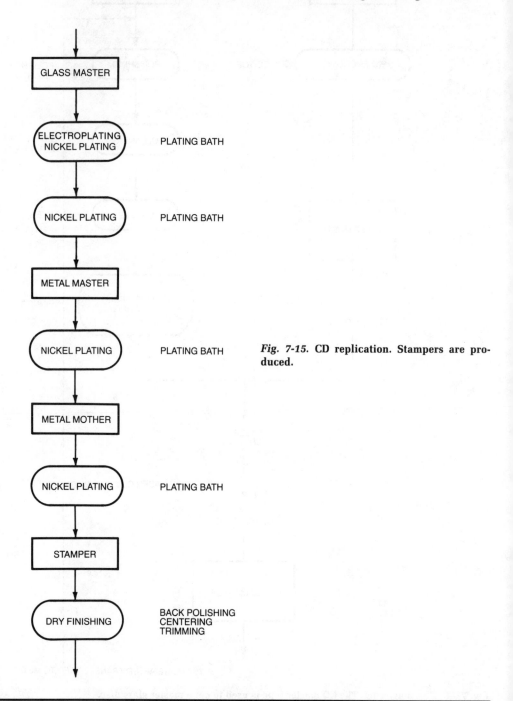

Fig. 7-15. **CD replication. Stampers are produced.**

By the same process, the resulting metal father can be used to generate a number of negative nickel stampers.

Mass production of discs can be accomplished with a number of different techniques. In one method, the final discs are produced with injection molding, as shown in Fig. 7-16. A polycarbonate material is used chiefly because of its low vapor absorption coefficient, about 70 percent less than that of a PMMA material. Polycarbonate material has an inferior birefringence specification especially when produced by injection molding; however, injection molding is a more efficient production method. After experimentation with different kinds of mold shapes and molding conditions, techniques for producing a single

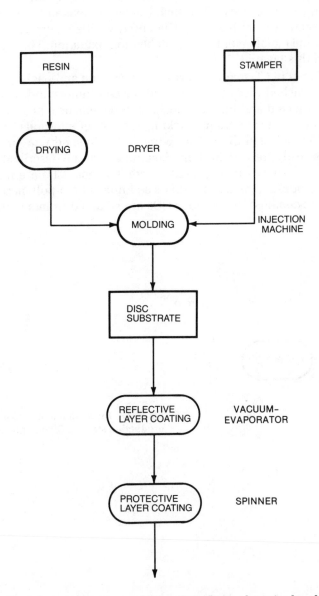

Fig. 7-16. **Disc manufacture. Discs are molded and a reflective layer is placed on the pit side.**

piece polycarbonate disc were achieved. Molten polycarbonate is injected into the mold cavity, and the disc (with pits) is produced; the process takes less than 15 seconds. The center hole is typically included simultaneously. In some methods the center hole is punched later, in a separate disc mold.

After molding, a layer of aluminum, silver, or gold (about 100 nanometers thick) is placed on the pit surface to provide reflectivity. The reflection coefficient of this layer including the polycarbonate substrate (note that the CD player laser must shine through the substrate to the metal layer) is specified to be between 70 and 90 percent. Aluminum evaporation is accomplished in a vacuum chamber, and takes about 15 minutes. Large racks of discs are treated in a batch process. Silver metalization can be accomplished through wet silvering, a continuous process. The metal layer is covered by an acrylic layer with a spin coating machine, and it is cured with an ultra-violet light. This layer protects the metal layer from scratches and oxidation. The label is printed directly upon this layer.

The final step in CD manufacturing is inspection and packaging, as shown in Fig. 7-17. Finished discs are inspected for continuous and random defects using both automated and human checking; birefringence, high frequency signal, frame error rate, noise, frame tracking, number of interpolations, and skew are measured on selected discs. Following packaging and wrapping, the Compact Discs are ready for distribution. A summary of the described disc manufacturing steps is shown in Fig. 7-18. Currently most CDs are manufactured with injection molding (as described) or a similar photo-polymerization technique. With new embossing methods CDs are stamped from a continuous roll

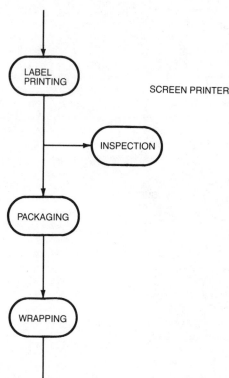

Fig. 7-17. Inspection and packing are the final production steps in CD manufacture.

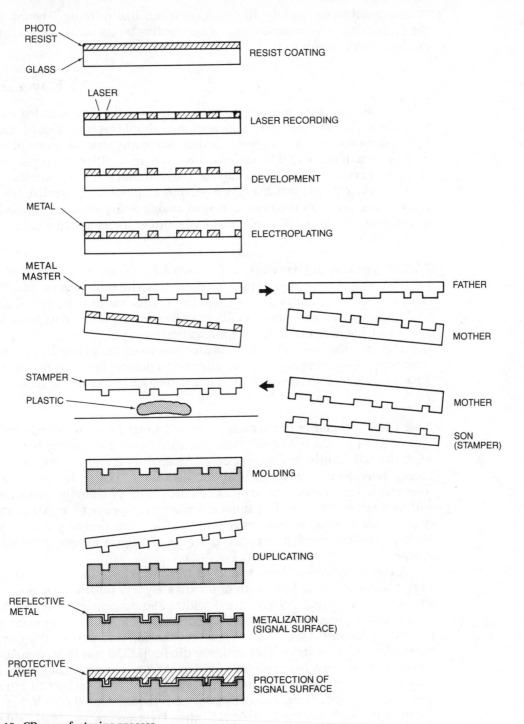

PHOTO RESIST
GLASS
RESIST COATING

LASER
LASER RECORDING

DEVELOPMENT

METAL
ELECTROPLATING

METAL MASTER
FATHER
MOTHER

STAMPER
PLASTIC
MOTHER
SON (STAMPER)

MOLDING

DUPLICATING

REFLECTIVE METAL
METALIZATION (SIGNAL SURFACE)

PROTECTIVE LAYER
PROTECTION OF SIGNAL SURFACE

Fig. 7-18. **CD manufacturing process.**

of ribbon substrate material in a process much like printing. The material is then sandwiched between transparent protective layers, along with a thin aluminized film for reflectivity.

7.3 Player Design

It is the Compact Disc player's function to recover the data encoded on Compact Discs. That job is more complicated than simple reproduction of the audio signal; demodulation and error correction processing must be accomplished as well as a complete digital to analog conversion. In addition, player sophistication is increased by the laser pickup used to read data and the automatic optical tracking and focusing systems which must be employed. The availability of CD players at a price competitive to that of analog audio reproducers for home, automotive, and portable applications constitutes a major engineering accomplishment.

7.3-1 Player Design Overview • A Compact Disc player uses a servo motor to rotate the disc, laser optics to read the data, demodulation, demultiplexing and error correction circuitry to assemble the digital audio data, and digital-to-analog converters, and low-pass filters to convert the audio data to an analog signal suitable for conventional amplification and application to loudspeakers. Controls and displays provide an interface between the player and the human operator. To control many of these subsystems, players incorporate one or more microprocessors into their design. A block diagram of a player is shown in Fig. 7-19.

7.3-2 Optical Pickup • The data is recovered from the Compact Disc with an optical pickup which moves across the surface of the rotating disc. A disc might contain 2 billion pits precisely arranged on a spiral track; the optical pickup must focus, track and read that data track. The entire lens structure, laser source and reader must be small enough to move laterally across the disc and float underneath the disc surface moving in response to tracking information or user random-access programming. Although design particulars vary among manufacturers, pickups are generally similar in design and operation. A typical pickup is illustrated in Fig. 7-20.

A conventional gas laser would be too bulky for a pickup; a CD pickup uses a semiconductor laser with approximately a 5 milliwatt optical output irradiating a coherent AlGaAs beam with a 790 nanometer wavelength. The light emitting properties of semiconductors have been utilized for many years. By adding forward bias to a PN junction the injected part of the carrier is recombined to emit light; light emitting diodes (LEDs) use this phenomenon. However, laser light is significantly different from ordinary light in that it is comprised of a single wavelength and is coherent with respect to phase, as shown in Fig. 7-21. Thus, a modified device is required, as shown in Fig. 7-22. A carrier is inserted in the GaAlAs activating layer, and recombined light emission is serially induced. However, the light must be amplified thus several steps are taken. Both sides of the activating layer are sandwiched within a material with a large band gap to enclose the carrier, and the refraction ratio at both boundaries of the activating layer is different to effect enclosure. Also, for amplification within the layer, the crystal surface in the direction of the light

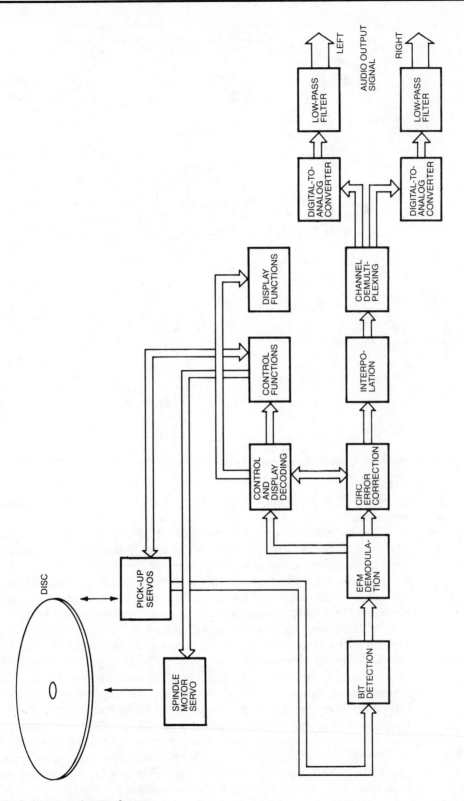

Fig. 7-19. **Block diagram of a CD player.**

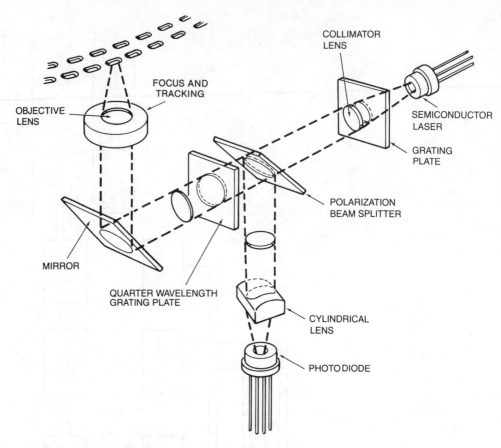

Fig. 7-20. **Optical pickup in a CD player. (Courtesy Sony)**

emission is made reflective and acts as a light resonator for constant wave emission.

The optical pickup is a sophisticated mechanism in several respects. Its optical design contains a diffraction grating, polarized beam splitter, quarter wavelength plate, and several lens systems, as shown in Fig. 7-23. The laser beam originates from the laser diode. A monitor photodiode is placed next to the laser diode to control power to the laser to compensate for temperature changes. The monitor diode conducts current proportionally to the laser's light output. If the monitor diode's current output is low with respect to a reference, current to the laser's drive transistors is increased to increase the laser's light output. Similarly if monitor current is too high, supply current to the laser is decreased to compensate. In a three-beam pickup the light from the laser point source passes through a diffraction grating. This is a screen with slits spaced only a few laser wavelengths apart. As the beam passes through the grating it diffracts at different angles, when the resulting collection is again focused it will appear as a bright center beam with successively less intense beams on either side, as shown in Fig. 7-24. The pickup system will use the center beam for reading data, tracking, and focusing. In a three beam pickup design, two secondary beams, the first order beams, are used for tracking.

The next piece of optics in the laser path lies at the heart of the pickup's

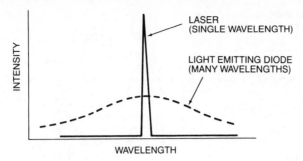

(A) Laser light contains one wavelength.

(B) Laser light is phase coherent.

Fig. 7-21. **Properties of laser light.**

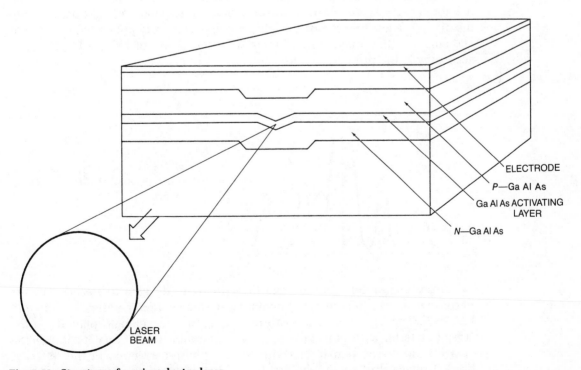

Fig. 7-22. **Structure of semiconductor laser.**

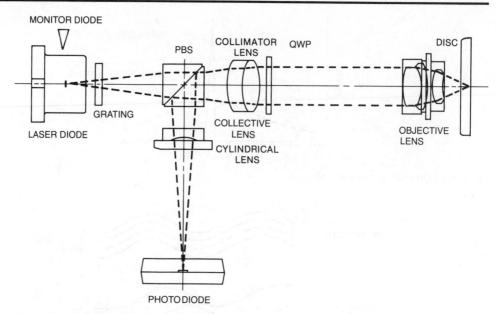

Fig. 7-23. **Construction of optical pickup in CD player.**

task of directing laser light to the disc surface, then angling the reflecting light to the photodiode. A polarization beam splitter (PBS) may be thought of as a one way mirror mounted at 45 degrees to the light path. For the light approaching the beam splitter, it acts like a window, but for the reflected beams it is a mirror redirecting the light by 90 degrees. The PBS is comprised of two prisms with a common 45 degree face with a dielectric membrane between them. That membrane acts as the one-way mirror; horizontally polarized light passes on to the disc surface while returning light, which is vertically polarized, is reflected and angled at the prism face. The relative intensity of the reflected light varies according to the presence or absence of pits on the disc.

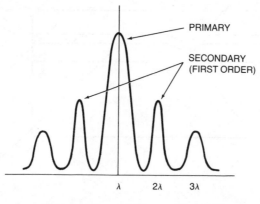

Fig. 7-24. **The diffraction grating splits the laser light into multiple beams.**

A collimator lens follows the PBS (in some designs it precedes it). Its purpose is to take the divergent light rays and make them parallel, as shown in Fig. 7-25. The light then passes through a quarter wavelength plate (QWP), a crystal material with astigmatic properties of double refractions. It controls the polarization of the light to accomplish the phase correction of the reflected light. A phase shift occurs the first time through the plate, then another phase

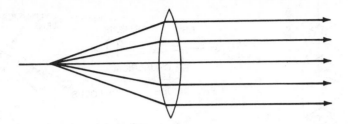

Fig. 7-25. **A collimator lens produces a beam of parallel rays.**

shift as the reflected light returns. The reflected light is polarized in a plane at a right angle relative to that of the incoming light thus allowing the PBS to properly deflect the reflected light.

The final piece of optics in the path to the disc is the objective lens; it is used to focus the light on the disc surface. The main spot is about 0.8 millimeter in diameter on the outer surface of the transparent polycarbonate layer. The refractive index of the polycarbonate layer is 1.55 and its thickness is about 1.2 millimeters so the spot is narrowed to 1.7 micrometers at the reflective surface, slightly wider than the pit width of 0.5 micrometer. The objective lens is attached to a two-axis actuator and servo system for up/down motion for focusing and lateral motion for tracking.

The data encoded on the disc now determines the fate of the laser light. When the spot strikes a land interval between two pits the light is almost totally reflected; when it strikes a pit (a bump from the reading side) with a depth (height) of about one-quarter the light wavelength in the transparent layer, the part of the beam reflected from the pit cancels the part reflected from the land, and little or no light is returned. Ultimately, a change in intensity will be deciphered as a 1 and unchanged intensity as 0. The varying intensity light returns through the objective lens, QWP (for another phase shift), the collimator lens, and strikes the angled surface of the PBS. The light is deflected 90 degrees and passes through a collective lens and cylindrical lens; these optics are used to direct the operation of the focusing servo system to keep the objective lens at the proper depth of focus. The beam's main function, however, is to carry the data via reflected light to a four quadrant photodiode. The electrical signals derived from that device will ultimately be decoded into an audio waveform.

7.3-3 Auto-Focus Design • To properly distinguish between reflective surface and pits, the laser beam must rely on interference in the reflected beam created by the height of the bumps, a mere 110 nanometer difference. Thus, the focus of the beam on the surface is critical; any unfocused condition would result in inaccurate data reading. Specifically, the laser must stay focused within a ±0.5 micrometer tolerance. Of course, even the flattest disc must contain deviations; specifically, disc specifications call for a tolerance of ±0.5 millimeter. Thus, the objective lens must be able to refocus as the disc surface deviates. This is accomplished by a servo driven auto-focus system, which utilizes the center laser beam, a four quadrant photodiode, control electronics, and a servo motor to move the objective lens. An operational diagram of the auto-focus system is shown in Fig. 7-26.

The unique property of astigmatism, creating distorted images, is used to

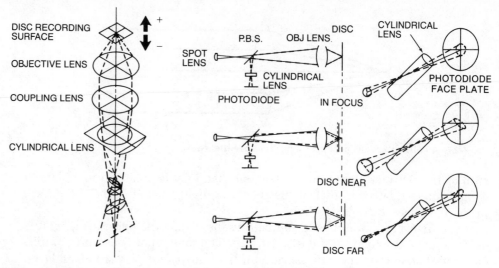

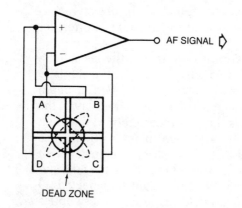

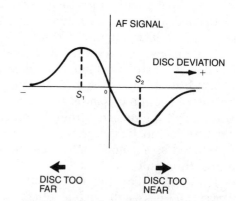

(A) The main beam in the optical pickup is used to generate an auto-focus signal.

(B) Astigmatism is used to trace an error pattern.

(C) The four quadrant photodiode converts the error pattern into an auto-focus signal.

(D) The AF signal shows disc position and is used to control a servo to maintain focus.

Fig. 7-26. **Principle of auto-focus signal detection.**

achieve auto focus in many CD player designs; the cylindrical lens just prior to the photodiode performs the essential trick needed to detect an out-of-focus condition. As the distance between the objective lens and the disc reflective surface varies, the focal point of the system also changes, and the image projected by the cylindrical lens changes its shape. The change in the image on the photodiode generates the focus correction signal.

When the disc surface lies at the focal point of the objective lens the reflected image through the intermediate convex lens and the cylindrical lens is unaffected by the astigmatism of the cylindrical lens and a circular spot strikes the center of the photodiode. When the distance between the disc and the objective lens decreases, the focal points of the objective lens, convex lens and the cylindrical lens move further from the cylindrical lens and the pattern becomes elliptical. Similarly, when the distance between the disc and the objective lens increases, the focal points are closer to the lens, and an elliptical pattern again results, but rotated 90 degrees from the first elliptical pattern.

The four quadrant photodiode reads an intensity level from each of the quadrants to generate four voltages. If a focus correction signal is mathematically created to be quadrants $(A+C)-(B+D)$, the output error voltage is a bipolar S curve, centered about zero. Its value is zero when the beam is precisely focused on the disc, and a positive-going focus correction signal is generated as the disc draws too near, and a negative-going focus correction signal is generated as the disc moves away. As in any closed loop system, such as the phase lock loop systems which keep motor speeds constant, the difference signal continually corrects the mechanism to strive for a zero difference signal and thus, in this case, a focused laser beam.

A servo system is used to move the objective lens up and down and thus maintain a depth of focus within tolerance. An electronic circuit is needed to decipher the focus correction signal and generate a servo control voltage; this circuit uses comparators and amplifiers to generate this servo voltage which in turn controls the actuator to move the objective lens. An example of this kind of circuit is shown in Fig. 7-27.

The objective lens is displaced in the direction of its optical axis by a coil and permanent magnet structure similar to that used in a loudspeaker except that the objective lens takes the place of the speaker cone. A two-axis actuator is used to accomplish this, as illustrated in Fig. 7-28. The top assembly of the pickup is mounted on a base with a circular magnet ringing it. A circular yoke supports a bobbin with both the focus and tracking coils inside. Control voltages from the focus drive circuit are applied to the bobbin focus coil and it moves up and down with respect to the magnet. The objective lens thus maintains its proper depth of focus. The other axis of movement, from side to side, is used to achieve tracking accuracy.

7.3-4 Auto-Tracking Design • On a Compact Disc modulated laser light carries the data and nothing (except light) touches the medium surface. That poses an interesting engineering challenge: how do we track a spiral pit sequence if there is nothing tangible to guide the pickup? The answer is the auto-tracking system found in all CD players.

The spiral pit track on a CD runs 30 revolutions within the width of a human hair, in addition, an off-center disc might exhibit track eccentricity of as much as 300 micrometers, and vibration further challenges the pickup's specification of tracking within a ± 0.1 micrometer tolerance. It is appropriate that a laser beam system is used for tracking; moreover, it would probably be impossible for any purely mechanical system to track as well. Two types of auto-tracking systems are used in CD pickups. The single beam system uses the center data beam for tracking (as well as data reading and auto focus), and the three beam system uses two additional beams solely for tracking. Both methods accomplish the same result with similar circuitry.

In a three beam pickup, the center beam is split by a diffraction grating to create a series of secondary beams of diminishing intensity. The two first order beams are conveyed to the disc surface along with the central beam, as shown in Fig. 7-29. The central beam spot covers the pit track while the two tracking beams are aligned above and below and offset to either side of the center beam, as shown in Fig. 7-30. During proper tracking, part of each tracking beam is

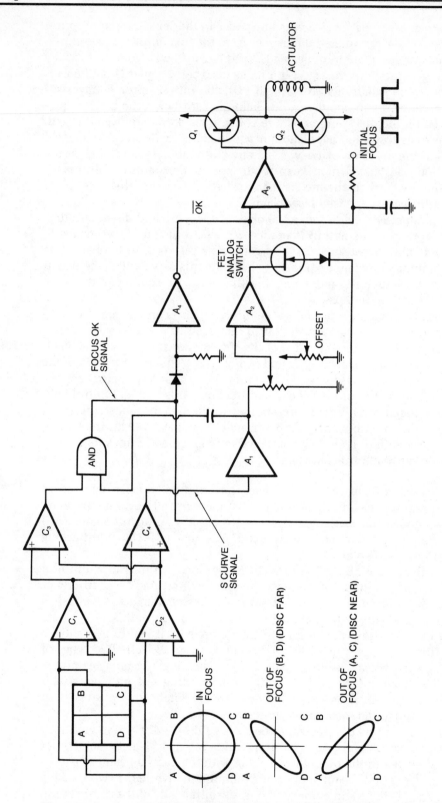

Fig. 7-27. Compact Disc auto-focus circuit. (Courtesy Hitachi Sales Corp. of America)

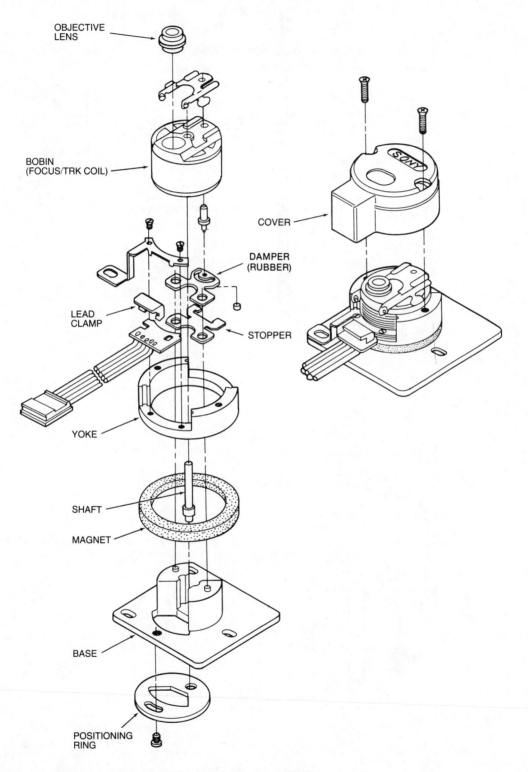

OBJECTIVE
LENS

BOBIN
(FOCUS/TRK COIL)

COVER

DAMPER
(RUBBER)

LEAD
CLAMP

STOPPER

YOKE

SHAFT

MAGNET

BASE

POSITIONING
RING

Fig. 7-28. **Compact Disc pickup two-axis actuator. (Courtesy Sony)**

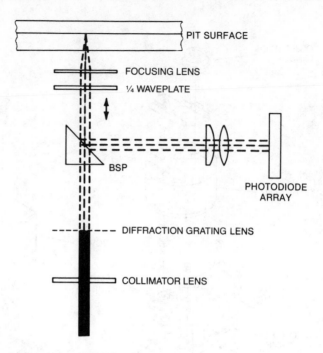

Fig. 7-29. Three beam tracking optical path.

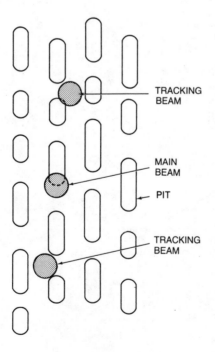

Fig. 7-30. Three beam tracking beam alignment.

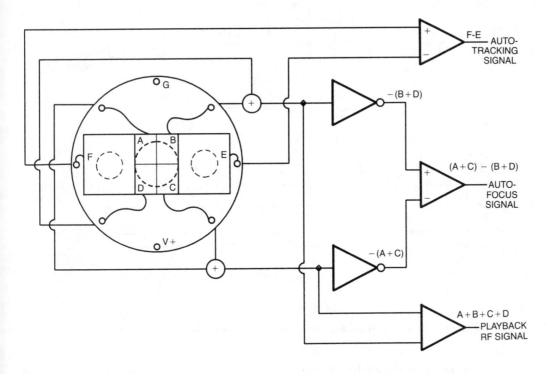

Fig. 7-31. **The four quadrant photodiode (A, B, C, D) is used for auto focus and data playback. Photodiodes E and F are used for auto tracking.**

focused on the pit circumference and the other part covers the mirrored area between pit tracks. The three beams are reflected back through the quarter wavelength plate and PBS; the main beam strikes the four quadrant photodiode and the two tracking beams strike two separate photodiodes mounted to either side of the main photodiode. The complete photodiode assembly for data, tracking and focusing is shown in Fig. 7-31.

As the three spots drift to either side of the pit track, the amount of light reflected from the tracking beams varies as one of the beams encounters more pit area which results in less average light intensity, while the other encounters less pit area and greater reflected intensity. The relative voltage outputs from the two tracking photodiodes thus form a tracking correction signal, as shown in Fig. 7-32.

Electronic circuits utilize the difference between the two signals for tracking correction. A delay circuit in the lead beam's signal path puts it on an equal basis with the following beam. If tracking is precisely aligned, the difference is zero. If the beams drift, a difference signal is generated, varying positively for a left drift and negatively for a right drift, to create an S curve tracking correc-

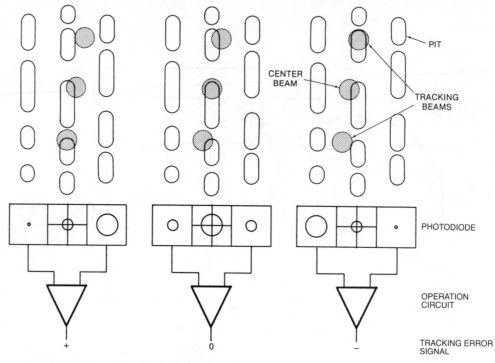

Fig. 7-32. **Generation of tracking error signal from two side beams.**

tion signal, as shown in Fig. 7-33. That signal is applied to the two-axis actuator assembly containing the permanent magnet and focus/tracking coil. To correct for a tracking error, the correction voltage is applied to the coil, the bobbin swings around a shaft to laterally move the objective lens so that the main laser spot is again centered, and the tracking correction signal is again zeroed.

A single beam auto-tracking system uses the same kind of closed loop to keep the pickup on track; however, instead of two secondary beams and outrigger photodiodes, the center beam, photodiodes, and a wedge lens are employed. As the beam drifts off track interference creates asymmetry in the beam. The wedge lens splits the beam into two parts; pairs of photodiodes detect the intensity difference in the two beams, and generate a correction signal, used to move the pick-up back on track through a servo system. To prevent offset problems in the correction signal, a low frequency alternating signal is applied to the tracking servo; the correction signal is thus modulated by a low frequency. When rectified, a drift-free correction signal is produced.

Aside from the tracking accuracy needed to keep the laser beam on track, a motor must properly move the pickup across the disc surface to track the entire pit sequence, and the pickup must be able to jump from one place on the disc to another, find the desired place on the spiral, and resume tracking. Both of these functions are handled by separate circuits, primarily using previously generated control signals. A coreless-type slide motor is used to provide constant tracking of the pit track. The tracking correction signal is used to provide control of the slide motor. Tracking in a CD player is thus similar to that of a conventional record player; in the same way that a record groove pulls

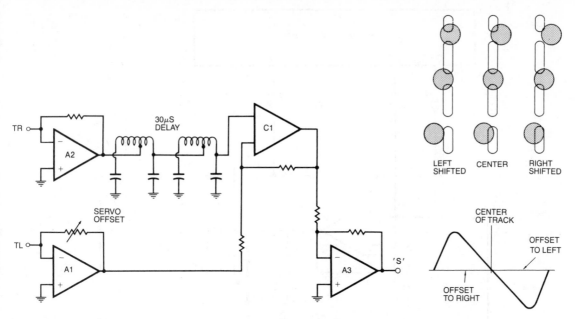

Fig. 7-33. **Generation of tracking "S" curve. (Courtesy Hitachi Sales Corp. of America)**

the stylus across an LP, the auto-tracking system pulls the pickup across a CD, keeping the pickup on track.

For fast forward or reverse the microprocessor takes control of the slide motor to provide faster motion than possible during normal tracking. When the correct location is reached, the S curve generated by the tracking correction signal is referenced to a microprocessor-generated control signal, and the circuitry is informed that proper tracking alignment is occurring. Just prior to alignment, a brake pulse is generated to compensate for the inertia of the pickup. The actuator comes to rest on the correct track, and normal auto tracking is resumed.

The reflectivity of discs may vary due to manufacturing process differences, soiling of the player optics, etc. It is important to maintain a constant voltage level for proper data recovery thus the gain of the output control amplifier is variable depending on the intensity of the laser beam. This gain adjustment is automatically accomplished during the initial reading of the disc table of contents and is maintained for the playing of the disc. This occurs under control of a microprocessor. In the design example shown in Fig. 7-34, the proper resistors are switched into the amplifier's circuit, varying gain about ± 10 dB. One of the control signals comes from the damage detection circuit which alerts the focus servo system to defective or damaged discs. In severe cases the objective lens is pulled away from the disc to prevent damage to the pickup.

7.3-5 EFM Demodulator • When the laser beam is reflected at the disc surface, the result is detected by a four quadrant photodiode. It is the voltage from this sensor which is ultimately transformed into the analog audio signal output from the player. However, important signal processing must first occur to properly convert the encoded data to an analog signal, as shown in Fig. 7-35. The signal encoded on the disc utilized EFM which specified that the

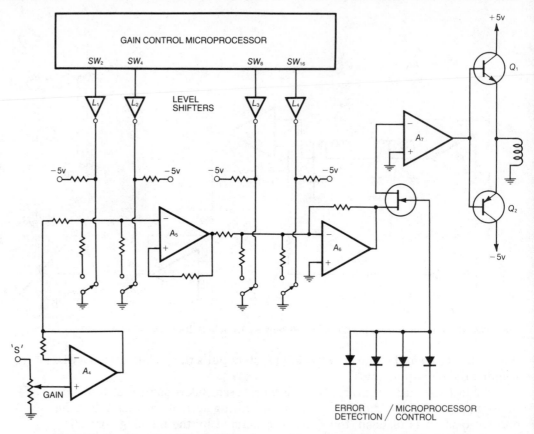

Fig. 7-34. **Tracking signal level shift and servo driver circuit. (Courtesy Hitachi Sales Corp. of America)**

signal be comprised of not less than 2 or more than 10 successive 0s between 0-to-1 transitions. This results in 9 different pit lengths from a pit length 3 channel bits long to a pit length 11 channel bits long. This sets the frequency limits of the EFM signal, since 3 channel bits correspond to a pit length of 0.833 micrometer and 11 channel bits correspond to 3.054 micrometers. This range is sometimes referred to as a 3T-11T signal with T referring to the period of one channel bit.

The photodiode and its processing circuits produce a signal resembling a series of high frequency sine waves called the EFM signal; it is sometimes referred to as an RF signal, or the eye pattern, in which the minimum time for 3T is approximately 700 nanoseconds. The information contained in the eye pattern is shown in Fig. 7-36. Whenever a player is tracking data, the eye pattern is always present, and the quality of the signal may be observed from the pattern. Although this signal is comprised of sine waves it is truly digital. It undergoes processing to convert it into a series of square waves more easily accepted by digital circuits; this does not affect the encoded data since it is the width of the EFM periods which hold the information. Prior to CD mastering, the Nonreturn to Zero (NRZ) signal on the master PCM video-based tape was converted to a Nonreturn to Zero Inverted (NRZI) signal in which the preceding

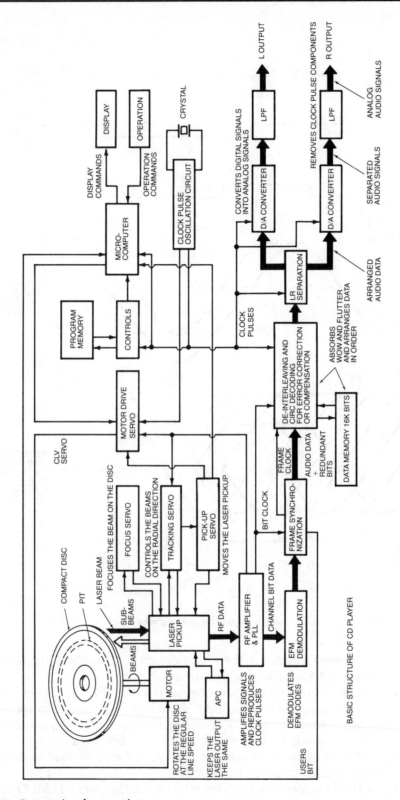

Fig. 7-35. **Output signal processing.**

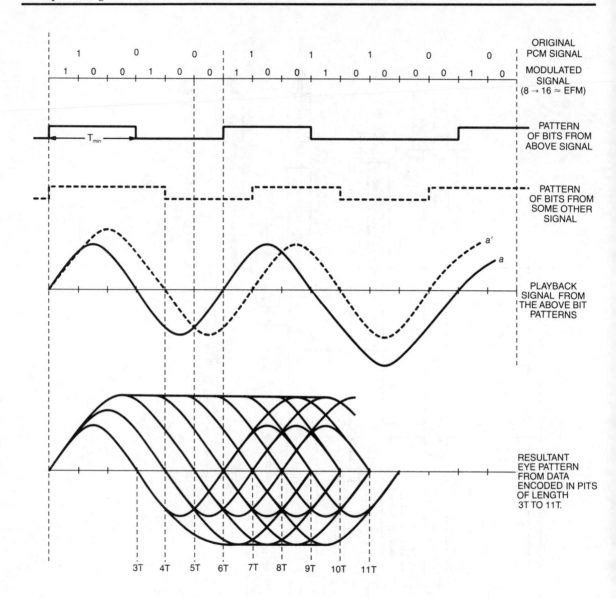

Fig. 7-36. **EFM encoding produces eye pattern (RF) signal output.**

polarity is reversed whenever there is a binary 1. The NRZI signal is now converted back to NRZ for further processing.

The first pieces of data to be recovered from the NRZ signal are synchronization words; frame synchronization bits which were added to each frame during encoding are extracted, symbol synchronization is generated to synchronize the 33 symbols of channel information in each frame, and the individual bits are used to generate a synchronization pulse to be used to determine whether the digital information is a 1 or a 0.

The EFM digital signal is now demodulated so that every 17 bit EFM word again becomes 8 bits. Demodulation is accomplished by logic circuitry or a

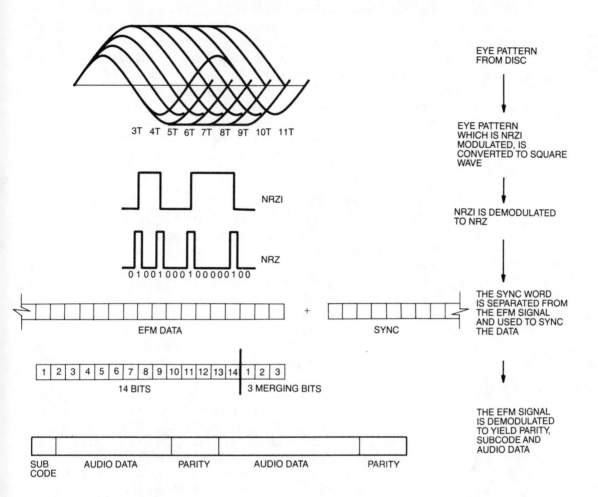

Fig. 7-37. **EFM demodulation of the eye pattern results in recovery of subcode, audio, and parity data.**

look-up table, that is, a list stored in memory which uses the recorded data to reference back to the original patterns of 8 bits. The process from eye pattern to demodulated data is summarized in Fig. 7-37.

During decoding, a buffer is used; disc rotational irregularities might make data input irregular but clocking ensures that the buffer output is precise. To guarantee that the buffer neither overflows nor underflows, a control signal is generated according to the buffer's current filled level and is used to control the disc rotating speed. By varying the rate of data from the disc, the buffer is maintained at 50 percent capacity.

7.3-6 Error Detection and Correction • Following demodulation, data is sent to a Cross-Interleave Reed-Solomon Code circuit for error detection and correction. The CIRC error correction decoding strategy uses a combination of two Reed-Solomon code decoders, C1 and C2. The CIRC is based on the use of parity bits and an interleaving of the digital audio samples. It enables complete correction of burst errors up to 4000 bits (a 2.5 millimeter section of pit track), and recovery through adjacent sample interpolation of losses of up to 12,300 bits (7.7 millimeter track length). The Two Reed-Solomon codes, C1 and C2, are used to accomplish correction. Code C1 is characterized by thirty-two 8 bit symbols, comprising 28 data symbols and 4 parity symbols. Code C2 is characterized by twenty-eight 8 bit symbols comprising 24 data symbols and 4 parity symbols. Together, in theory, two erroneous symbols or four erasures (symbols whose value is not known, but position is) can be corrected. Fig. 7-38 illustrates the operation of CIRC decoding. The C1 and C2 codes are shown, as well as delays which de-interleave the samples and permit more efficient error correction.

For maximally effective correction, decoding must be alternatively repeated as C1 then C2, accomplished twice (C1, C2, C1, C2). Two kinds of parity bits were added during encoding, the P and Q parity bits (not to be confused with P and Q subcode), both used solely for error detection. The Q parity adds four words of parity bits to each CD frame, during reproduction, Q parity bits are used to determine whether there was any error in the encoding process. The P parity bits are added after interleaving to detect more severe drop-outs; another four words of parity are added to each frame.

During reproduction, P parity is checked first. This process is called C1 decoding. The C1 decoder detects burst errors and corrects random errors. It cannot make erasure correction. The 32 symbols of each frame (24 audio symbols and 8 parity symbols) are input in parallel to the 32 inputs of the CIRC decoder. The delays at the input of the C1 decoder which separate the even and odd numbered symbols are all of duration equal to the duration of one symbol, so that the information of the even symbols of a frame is cross-interleaved with that of the odd symbols of the next frame. This enables the C1 decoder to correct small errors in adjoining symbols. The C1 decoder is designed according to the rules for a Reed-Solomon code with $n1 = 32$, $k1 = 28$, and $s = 8$ in which the 32 symbols are multiplied by the C1 parity check matrix to produce four syndromes. From these syndromes it fully corrects one erroneous symbol; in the event of multiple errors, it passes the symbols unchanged, attaching to each of its 28 output symbols (the four C1 parity symbols have been dropped) an erasure flag, indicating that each symbol may contain an error. If there is no flag attached to a symbol arriving at the C2 decoder, that symbol is assumed correct. If a defect is detected, a pulse is generated to denote the frequency of errors; the count should show less than 220 errors per second; more errors might require muting of the audio signal. For a count of less than 220, depending on severity of errors, different methods of error correction are possible, such as holding the previous word, linear interpolation, and higher order interpolation.

The delay lines prior to the C2 decoder are of an unequal duration and longer than those prior to the C1 decoder such that errors which occur in one word at the output of C1 are spread over a number of words at the input of C2. This reduces the number of errors in any one C2 word; thus better enabling the C2 decoder to correct burst errors.

In C2 decoding, Q parity is detected and corrected. The C2 decoder is designed according to the rules for a Reed-Solomon code with $n2 = 28$, $k2 = 24$, and $s = 8$. It corrects burst errors and any random errors which the C1 decoder was unable to correct. Specifically, it can correct a maximum of four

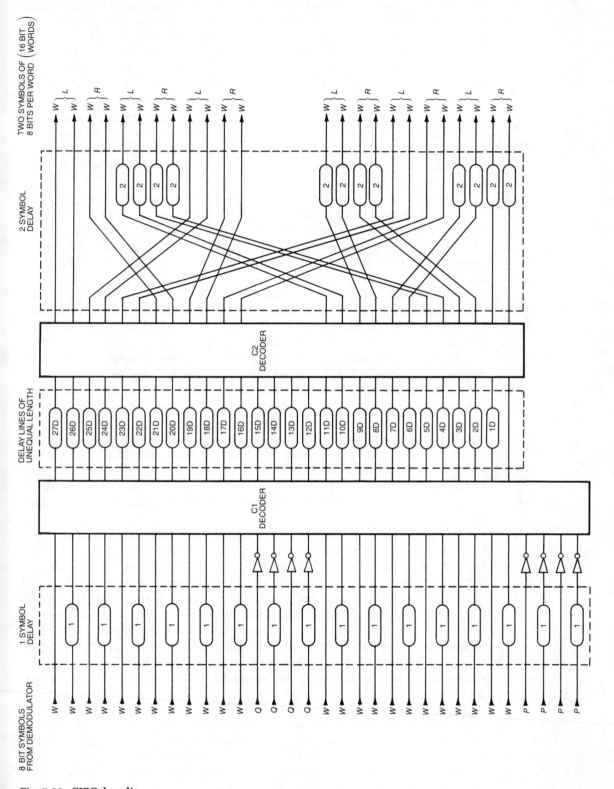

Fig. 7-38. CIRC decoding.

errors with an erasure-position method. If there are more than four errors, and the C2 decoder cannot correct all errors, it outputs the 24 data symbols (the four C2 parity symbols have been dropped) uncorrected, but marked with an erasure flag. Most of these symbols are usually error free and can be reconstructed by linear interpolation since the combination of the C1 and C2 flags can be used to aid interpolation. If an error is detected as a result of C2 decoding, a pulse is generated; the count should be less than approximately 100 counts per second for these errors. Double error correction is performed in C1 decoding and error pointers are set to avoid miscorrection; these results are checked in decoder C2 by comparing these pointers with the location found by decoding. Erasure correction may be accomplished in the C2 decoder by using pointers set in the C1 decoder.

Theoretically, the raw bit error rate on a CD is between 10^{-5} and 10^{-6}, that is, there is one incorrectly recorded bit for every 10^5 (100,000) to 10^6 (1 million) bits on a disc. Following CIRC error correction, the bit error rate is reduced to 10^{-10} or 10^{-11}, or less than one bad bit in 10 billion. In practice, because of the data density, even a mildly defective disc can exhibit a much higher bit error rate.

Interpolation and muting circuits follow the CIRC decoder in a Compact Disc player. The function of these error concealment circuits is to process errors which have been detected, but not corrected by the CIRC decoder. The aim is to reduce such errors to inaudibility. Only uncorrected symbols, marked with erasure flags, are processed. All valid audio data passes through the concealment circuitry unaffected, except in the case of data surrounding a mute point, which is attenuated to minimize audibility of the mute. Concealment methods vary according to the degree of error encountered, and from player to player. In its simplest form, when a single sample is flagged between two correct samples, mean value interpolation is used to replace the erroneous sample. For longer consecutive errors, the last valid sample value is held, then the mean value is taken between the final held value, and the next sample value. The maximum number of erroneous symbols which leave the decoder uncorrected, but can be concealed by linear interpolation between adjacent sample values is 12,300 audio data bits.

If large numbers of adjacent samples are flagged, the concealment circuitry performs muting. Using delay lines, a number of previous valid samples (perhaps 30) are gradually attenuated with a cosine function to avoid the introduction of high frequency components, gain is kept at zero for the duration of the error, then gain is gradually restored. Errors which have escaped the CIRC decoder without being flagged would not be detected by the concealment circuitry, and thus would not undergo any concealment; they might produce an audible click in the audio reproduction. Not all CD players are alike in terms of error protection. Any CD player's error protection ability is limited to the success of the strategy chosen to decode the CIRC data on the disc, and perform concealment.

7.3-7 Output Processing • Following error correction, the digital data is processed to recover control information, and demultiplexing. During encoding, eight bits of subcode control information per frame were placed in the bit stream. During decoding, control data from 98 frames are read and placed together to form one block, then assigned eight different channels to provide control information. The P channel designates lead-in, lead-out, and play areas of the disc. Binary 1s denote lead-in areas, 0s denote play areas, and alternating 1s and 0s denote lead-out areas. The Q channel contains track numbers, index numbers, elapsed time within a track in minutes, seconds, and frames, and elapsed time since the first music track.

The 16 bit audio data is demultiplexed so that simultaneous data may be presented to the D/A converters and output filters in the same sequence and at the same rate as they were originally recorded. Following this processing, the data is converted into twos complement or offset binary, depending on the D/A converter used in the player. In some players, only one D/A converter is used; left/right data words are alternately converted, then the analog signals are demultiplexed. Care must be taken to compensate for the 11.34 microsecond delay between channels.

The output stage of a Compact Disc player contains digital to analog converters, output sample and hold circuits, low-pass filters, and output amplifiers. All of these circuits have been described in detail in Chapter 4.

7.3-8 System Interfacing and User Information • The Compact Disc system is the first digital audio consumer product thus it presents new questions to the user. There are several elementary considerations important in selecting a player, integrating it with a playback system, and using it to its best advantage. While the theory of de-digitization is identical for all players, the actual implementation of these circuits dictates differences in performance and overall quality as evident in individual players. There is great variability in factors such as stability of tracking, and performance of error protection. Players differ in the perceived quality of the sound itself. This is largely due to the design decisions embodied in the output circuitry of a particular player. For example, the quality of the D/A converter plays a large role, in addition, while most players use two converters, one for each output channel, other players make-do with one. Manufacturers particularly differ on the type of output filter; some employ conventional brickwall analog filters, while others recommend digital oversampling filters followed by gentle roll-off analog filters.

The Compact Disc player is designed to interface with existing analog audio reproduction equipment, and it points toward the not-so-distant future when digital audio tape recorders, digital preamplifiers and power amplifiers, and erasable CD systems will become available. For now, the CD represents many major improvements over the LP in sound reproduction: longer playing time, negligible wow and flutter, rumble, coloration, noise, no pickup mistracking, low distortion, large dynamic range, large frequency response, high channel separation, mechanical shock resistant, acoustic feedback resistant, few alignments or adjustments, phase linearity, no surface noise, few disc handling problems, no disc wear, or aging—It is, in short, a near-ideal audio medium.

Given the high performance specifications of a CD system, what is to be expected when it is interfaced to a conventional analog stereo system? A paradox might present itself. Given a superior source, the output should be superior. This is true in many respects; for example, no rumble, clicks or pops, mis-tracking, etc. However, a playback system which sounds fine with an LP source might sound inferior with a CD source. But the problem is with the system, not the CD. The digital source places unprecedented sonic demands on a playback system in terms of frequency response, dynamic range, etc. That might reveal previously overlooked weaknesses in the system. For example, the increased dynamic range of a Compact Disc might overload an amplifier, particularly during peak passages. Given that a CD has 10 to 20 dB more dynamic range than an LP (a conservative estimate), consider this example: an amplifier producing an average of 1 watt would require 10 watts to handle a 10 dB musical peak from an LP. But a peak with an additional 10 dB of level from a CD would demand 100 watts from the amplifier. If the power was not there, clipping distortion would result. The choice is to lower the overall level with the existing amplifier, or buy a higher-powered one.

A similar problem is encountered with loudspeakers. The extended frequency response from a CD places new demands on tweeters and woofers; only high quality loudspeakers are able to faithfully deliver the information contained on a CD. Also, because of the larger dynamic range of a CD, and the resultant increased output from an amplifier, a loudspeaker must have a greater power handling capacity to prevent self destruction. In addition, if the amplifier is delivering a distorted signal, this further accentuates the abuse to the loudspeaker. The problem of amplifier power demands outlined previously could be solved with the selection of high efficiency loudspeakers. Efficiency varies over a factor of 10; switching from a 1 percent efficient loudspeaker (typical) to a 2 percent efficient loudspeaker would effectively double the power output of the amplifier.

The Compact Disc thus raises the standard required of a sound reproduction system. Higher quality components are needed to take advantage of the CD's source fidelity, and avoid a fidelity mismatch which could expose other system weaknesses. In general, under most listening conditions, a good quality analog reproduction system is adequate for a CD source. Of course, the other advantages of the CD such as longevity always remain. With the advent of digital amplifiers and new loudspeaker designs, reproduction system standards will more generally match those of the Compact Disc.

To maintain the high fidelity standards of a Compact Disc, the discs themselves must be maintained. While discs are largely impervious to dust and dirt, the bane of analog recordings, they are not indestructible. They should be handled with the same regard as an LP, particularly to prevent scratching either the data or label side of the disc. Superficial abuse to the disc, such as dirt and fingerprints, can be wiped from the disc surface; it is recommended that a disc be wiped radially, that is, across center to outer edge, rather than around its circumference to avoid putting scratches in line with the pit spiral, a condition which could make error correction difficult.

Of course, the sound from a Compact Disc player is only as good as that recorded on the disc itself. An older analog master recording issued on a CD format cannot approach the quality of the digital medium hence the reproduced sound is affected by the limitations of the master recording itself. Noise floor, distortion, limited dynamic range, and frequency response are all factors. The solution is the purchase of recordings originally recorded digitally which preserve the fidelity of the performance. Many CDs now carry a three-letter code which tells the history of the recording; the three letters represent: session recording, mixing-down/editing, and finished product format. An A represents an analog stage, while D is digital. Thus, an ADD code describes an original analog recording mixed down to a digital recorder, and mastered to Compact Disc. A DDD disc would be all-digital.

The CD also places great demands on the talents of recording engineers. Many techniques suitable for analog are unsatisfactory for digital recordings. For example, many microphones have been designed with a built-in high frequency boost to help high frequencies survive the analog signal chain. Such microphones, particularly when used in a close-microphone application, can result in a strident, unnatural sound in a digital recording. Recording engineers and other audio technicians must learn to meet the opportunity and challenge of the new, more demanding medium.

7.4 Compact Disc Read Only Memory

The Compact Disc is a remarkable digital storage medium. As we have seen, over 5 billion bits of audio data may be reliably stored on a disc thanks to EFM modulation and CIRC error correction techniques. However, that medium is available for other types of data not restricted to audio applications. In place of audio data, computer software or other published material may be stored in a read-only format, and delivered as a video signal as opposed to an audio signal. For example, over 275,000 pages of text, each holding 2000 characters, could be placed on one Compact Disc and displayed on a television monitor. Furthermore, the storage potentials of an audio-only CD are not exhausted; using the subcode bits in each frame, limited amounts of video information may be stored on audio CDs.

7.4-1 Compact Disc Subcode • An audio Compact Disc contains unused data capacity in the guise of the subcode bits. The eight bit subcode of user bits is contained in every frame and are designated as P, Q, R, S, T, U, V, and W, sometimes referred to as the PQ code. Only the P and Q bits are used in the audio format, and contain information such as the total number of selections on the disc, their beginning and ending points, index points within selections, preemphasis on/off, and end point of the disc. The other six bits which account for about 20 megabytes (8-bit bytes) of storage, are available for video information. Since the number of bits available in each frame is small, the entire number of subcode bits available, over 98 frames, is collected to form a subcode block, complete with its own synchronization word, data, and parity, as shown in Fig. 7-39. For a video image, the R, S, T, U, V, and W data is collected over thousands of blocks to form an image. A CD holding an hour of audio data can hold up to 700 still video images, drawing a new image from the disc every five seconds.

The possibilities are varied for such a CD format; using an I/O port on the CD player an adaptor connects the player to a television, and the images are viewed as the music is reproduced. Still pictures relating directly or conceptually to the audio material are displayed, as are liner notes, lyrics, or other alphanumeric information. Either video camera images or computer generated still images can be stored on the music disc.

7.4-2 CD-ROM • Compact Disc Read Only Memory (CD-ROM) is the logical transition toward the marrying of digital audio and information systems. Rather than store music, the CD format can be treated as a read only memory system, used for any kind of program material. Apart from modulation and error correction overhead, an audio Compact Disc holds a maximum of 6.3 billion bits, or 782 megabytes (8-bit bytes), of user information (1.41 million audio bits per

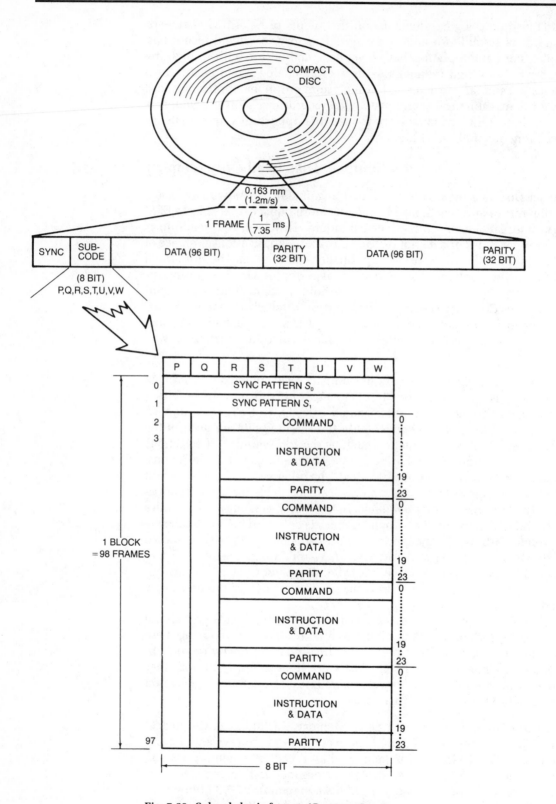

Fig. 7-39. **Subcode basic format. (Courtesy Sony)**

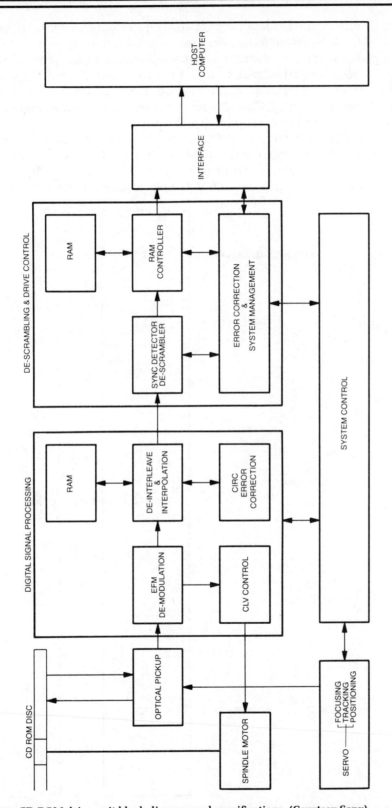

Fig. 7-40. CD-ROM drive unit block diagram and specifications. (Courtesy Sony)

second, 74 minutes). This large storage area, equivalent to 1500 half-megabyte floppy disks, can be given to the storage of other information, such as computer applications software, operating systems, on-line data bases, published reference materials, directories, back issues of journals, encyclopedias, libraries of still pictures, or other type of information not requiring frequent updating. In addition to mere storage, a CD-ROM system is interactive, for example, allowing the user to index and cross-reference the information.

As currently available, a CD-ROM stores between 550 and 650 megabytes of data with a special format; a CD-ROM disc automatically identifies itself as differing from an audio CD. Bit error rates can be lowered to 10^{-12} or less because of extended error correction data held in an auxiliary data field. The

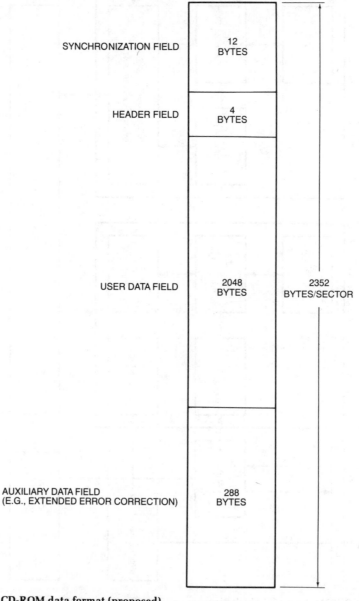

SYNCHRONIZATION FIELD — 12 BYTES

HEADER FIELD — 4 BYTES

USER DATA FIELD — 2048 BYTES

2352 BYTES/SECTOR

AUXILIARY DATA FIELD (E.G., EXTENDED ERROR CORRECTION) — 288 BYTES

Fig. 7-41. **CD-ROM data format (proposed).**

modified player contains laser optics, modulation, and error correction, but the audio output section is replaced with a computer interface to accommodate the ROM data. A block diagram and specifications for a typical CD-ROM drive unit are shown in Fig. 7-40.

The currently available CD-ROM data format provides for a maximum of 333,000 data sectors, each sector containing a synchronization, header, user, and auxiliary field, as shown in Fig. 7-41. The CD-ROM sectors are addressed by playing time in minutes, seconds, and sector number; this address is contained as a 3 byte binary coded decimal number in the header field. A sector address of 73-15-60 denotes the 60th sector in the 15th second of the 73rd minute on the disc. Given 2048 bytes per sector, and 333,000 sectors per 74 minute disc, a CD-ROM disc would hold an absolute maximum of 682 megabytes or user digital information. Data rate is limited to 1.41 megabits per second using components found in audio players; however, proprietary ROM systems can achieve faster rates with special components. Access time for any point on the disc is less than a second. A CD-ROM disc could hold 275,000 pages of alphanumerics, or 18,000 pieces of computer graphics, or 3600 still video pictures.

In other words, the CD-ROM would form the basis for a new electronic publishing medium applicable to book publishing, dictionaries, technical manuals, business catalogs, computer storage, expert systems, artificial intelligence, and so on. The Compact Disc is thus more than a replacement for the vinyl LP record. It represents an entirely new technology of information dissemination.

7.4-3 CD-I • The Compact Disc Interactive (CD-I) format extends the applications of the Compact Disc to new and diverse markets. CD-I is a product-specific application of the CD-ROM format. Rather than store specific data on a CD-ROM, or music on an audio CD, CD-I permits storage of a simultaneous combination of audio, video, graphics, text, and data, all functioning in an interactive manner. CD-I is thus a multimedia extension of CD-ROM. Because CD-I players also reproduce conventional audio CD discs, CD-I is also an enhanced CD audio system.

The CD-I format defines both hardware and software standards, and integration of functions. To provide flexibility in its intended applications, CD-I permits different levels of audio fidelity level, video resolution, and color coding. For example, adaptive delta pulse code modulation (ADPCM) can be used to encode audio data. This is a variation of linear PCM; because of its ability to store digital audio data with fewer bits it is extremely effective for CD-I, which must share disc space with other kinds of data. Using 8- or 4-bit ADPCM, and different sampling rates, audio quality levels approximating the quality of LP records, FM broadcasts, AM broadcasts and telephone lines can be obtained. Linear PCM can be used for CD quality audio.

Longer CD-I playing times are achieved by dividing the disc into channels, and executable object code provides for real time interactivity. Physical interleaving of data ensures synchronized presentation of different data types. Different CD-I data forms permit maximum bandwidth storage of data that degrades gracefully (eg. audio/video), and extended error detection and correction for data that does not (eg. text/binary).

A single CD-I disc might contain recorded music or speech, color pictures, animation, graphics, text, or any combination of these. For example, a CD-I dictionary might contain a word and its definition, as well as spoken pronunciation, pictures, and additional cataloging to synonyms, antonyms, word relationships, origins, or translations into foreign languages. Multimedia, interactive fiction works could be provided with a labyrinth of plot deviations, differing each time the disc is read, perhaps steered at the discretion of the reader. The CD-I format thus presents considerable opportunities for the hardware, software, and publishing industries to provide consumers with new forms of interactive entertainment and education.

7.4-4 CD-V • Compact Disc Video (CD-V) merges Compact Disc audio with Laservision video. CD-V players are able to play 8-inch and 12-inch diameter video discs with digital audio soundtracks. The 8-inch disc offers two 20 minute sides, and the 12-inch disc offers two 60 minute sides. In addition, CD-V players accommodate regular-size CD discs with both video and digital audio, recorded on one side only. Called CD-V Singles, these discs store about 5 minutes of high quality video and digital audio, with an additional 20 minutes of digital audio only. Because the digital audio data is contained on the innermost disc diameter, the audio portion of these CD-V Singles are also playable on regular CD audio players (of course, the video signal cannot be played). The video portion is stored on the outermost disc diameter, in analog form. CD-V players are also able to play audio-only CDs.

Because television standards differ around the world, CD-V players and discs are manufactured according to both the NTSC (U.S. and Japan) and PAL/SECAM (Europe) standards. These two standards are incompatible. To visually differentiate CD-V Single discs from audio CDs, CD-V Singles are treated with a special lacquer to give them a gold hue, as opposed to the silver color of audio CDs. With CD-V, consumers may enjoy the highest quality video and audio, together.

Chapter 8

A New Beginning

Introduction

Philosopher Martin Heidegger suggested that a science's level of development may be measured by the extent to which it is willing to undergo change in its fundamental concepts. If that supposition is true, we are indeed privileged to live in an era of technological sophistication. The dramatic changes experienced in the twentieth century which have accelerated with each next decade have demonstrated science's readiness to carry technology increasingly beyond the bounds of expectation. Of course, science is only a tool of society thus the technological revolution mirrors our own willingness to advance the sophistication and complexity of our lives.

8.1 Music and Technology

Today change is everywhere. Everything old is challenged and often redefined. Ways of thinking adapt to the new ways of creating. Barriers and identities shift and give way, extremes come together. The science of sound and the appreciation of music have combined to form the audio technology industry. And because of that industry the experiences of live and recorded music are drawing closer. The sophistication of digital audio's reproduction blurs the distinction between the reality of performance and the reality of re-created performance.

Digital audio technology is not an extension of existing audio technology. It is a radically new approach. As we have seen, the hardware recording and reproduction equipment is different, the recording media are different, even on the most basic conceptual levels, digital audio is unlike analog audio. The 100 year evolution of analog audio has reached maturity at precisely a pivotal point in our overall technological evolution. Computers and associated digital technology is increasingly influencing everything we do including, of course, the recording and reproduction of music.

8.1-1 The End of Analog • Surely history will always marvel at pictures of water wheels, wind-up clocks, steam engines, and analog audio equipment as quaint and clever inventions. Wax cylinders, rubber, wax, acetate, shellac, vinyl discs, wire recorders, open reels, cassettes, 45, 33⅓, 78 rpm, 1⅞, 3¾, 7½, 15, 30, and 60 ips, 2, 3, 4, 8, 16, and 24 tracks, ⅛, ¼, ½, 1, and 2 inch tape widths, RIAA, NAB, and IEC standards—that entire chronology of analog development has reached the end of its course, made obsolete by newer developments.

Granted the phonograph was hailed as the greatest invention in an age of great inventions, and development of audio technology was swift and remarkable; any new technology makes tremendous gains initially, then the pace of improvement slows as the technology reaches maturity. But with analog audio its very nature determined its life span; with analog the limitation of using a medium in which noise was inherently indistinguishable from signal dictated a finite evolution. The analog phonograph record was a self limiting, fragile medium; even the best components could only more accurately convey the pops and clicks from the dust and dirt slowly settling into the storage places of the recordings. And even at best, a new analog recording compared to the live music was like a writer's analogy and his attempt to substitute a well-turned phrase for something indescribable.

Analog music recording was time spent in waiting until the time when a more sophisticated technology came along. Was there ever really any comparison between the experience of live music, and the pressure wave resulting from the LP? After lending an ear to live music, then listening to analog playback, could anyone ever deny that the reproduction was lacking? Analog recordings never fooled anyone, not one bit. They were only recordings, generations and generations removed from their live birthright. Even in the place of origin, the recording studio, it was a battle against loss of fidelity, and restriction of creative impulse. In any control room, the performers on the other side of the glass were only reminders of the imperfect reproduction being accomplished.

8.1-2 Higher Fidelity • Now the wait is over. With digital music one can at last listen to playback and begin to feel as if one is there—at the performance. High fidelity will have to be redefined as higher fidelity. Moreover, since we have gone digital, our information is irrevocably permanent. Future systems of still higher fidelity will be devised but our digital data will never age; perhaps future reproducers will even interpolate that data to provide new aspects of information from its content. The Compact Disc is replacing the LP as our most popular storage medium. For information storage, analog technology was doomed, whereas digital certainly is not. Digital is an inherently more suitable design method for audio. Design is essentially a question of technology, but with music it is ultimately a question of hearing. The limitations of analog storage are simply too audible. In the past we could never satisfactorily listen to recordings because they were analog.

Our conversion to digital audio will necessitate a complete re-thinking of the recording process. Because of digital technology, the amount of information is so much more vast, and our methods of analyzing and processing it so much more efficient, because the power of the available music information is much greater, we should speak in terms of a digital re-creation of live music. That is truly how much information we have now, enough to re-create an acoustic event, instead of merely record it.

With digital technology the nature of the storage medium and its content have been divorced. With binary coding we can store information as pure data inherently distinguishable from the circumstances of its storage. Information has been provided the ideal storage method. And one of the most pleasant kinds of information, long in need of good storage, is music. Ironically, now that music storage is no longer a design problem, the next weaker links in the chain now present themselves. Both phases of acoustical/electrical conversion are the new limiting factors of music recording and reproduction. Digital audio is a breakthrough and a time for new questioning....

8.2 The Battle of Digital Audio

Some people have stridently challenged the digital technology currently being used in audio equipment; everything from supposed lack of reverberation to granulation noise has been critiqued. Of course, not all such questions are valid, but the questioning motive is. Our discussions have shown that the technology now being used in digital audio equipment is relatively primitive. It is as relatively straight-forward as the wax cylinder recorder/reproducer and its hand crank. It is a new technology, and that dictates simplicity. Of course, a Compact Disc player represents a more sophisticated technology, but in future historical terms, today's digital technology is very limited in its ability. Deservedly, current digital audio technology bears the burden of serious questioning.

The question of sampling rate is very difficult. Sampling theory argues for a rate at twice the highest throughput frequency—40 kHz for a 20 kHz signal. But aliasing is a potential for severe distortion; we require steep filters to band limit the signal. But what about the resultant phase shifts? Do we try to compensate for that with time correction, raise the sampling rate to allow for less radical filter characteristics, or devise new techniques such as oversampling? Moreover, is 20 kHz itself high enough for our enjoyment of music? Sooner or later every question is answered, denied as being invalid, or answered with a solution which represents an improvement. This process clears the way for new questions.

Similarly, quantization criteria raise questions and crowds technology's ability. Sixteen bits and its 65 K levels of resolution seem to provide for a reasonable dynamic range with reasonably low granulation noise, and it is adopted as a standard. For now, the technology appears solid enough to justify the start of production. However, every manufacturer has already considered 16 bit ranging systems or fixed 18 bits. But that is still to come; the technology is too expensive now; moreover, the research and development expense for 16 bits needs to be recouped. In other words, expediency and economics is what technological evolution is all about. Meanwhile, controversy rages—what sampling rate, how many bits, if digital has a "sound," when is phase shift audible, etc. These questions serve to further the technology; on the other hand, paradoxically technology seemingly furthers itself, as far as all of these questions— they don't make much difference. What we buy now will be superseded by something better.

8.2-1 The Stress of Change • If there is any truism in our world, it is this— technology will evolve. Our digital audio systems are only prototypes of what is to come. Techniques such as PCM and delta modulation will undoubtedly reign, then abdicate to newer technology, technology already being readied in research labs. For example, even before the Compact Disc had been legitimately introduced in the United States marketplace, Philips had already announced work on an erasable CD.

Ironically, that fact creates trouble for the near term. Part of analog's success can be attributed to its adaptability; it is a very flexible medium and one given to change. For example, a single groove can accommodate one, two, or four audio channels. But with digital technology the specification is more precise, and constraining. To properly anticipate future change, we must attempt to build it into today's specifications. Decisions of sampling rate, word length, format, error detection and correction must serve their purpose before we again find justification to announce obsolescence. A part of a technology's success is measured in its longevity. No matter how many CDs are sold in the next 15 years, if an incompatible super CD is then marketed, we must conclude that the original CD was a failure because it was unable to accommodate change.

Yet to effectively compete in an increasingly sophisticated entertainment industry, the music industry must offer a newer and better product. Likewise, competition internal to the industry provokes all manufacturers to constantly revise. Economics plays an important role; much of today's research and development is being devoted to digital devices because they offer the most cost effective way to provide a perceptively high quality product. With digital, that design process is facilitated. But will the economic incentive to provide newer and better products result in too-rapid obsolescence?

8.2-2 The Integrated Advantage • The future home reproduction system will be a sophisticated retrieval system, perhaps networked to audio libraries. The dissemination medium will be digital, thus permitting the consumer to audition a signal identical to that heard in the control room. It is in the consumer market that digital technology's impact on the audio industry will be most dramatic, and the battle of digital audio will be won in the home. Over the next several decades, analog playback and recording systems in homes will be displaced by new digital ones. Perhaps even more importantly will be the software impact as analog record and tape collections yield to digital storage on Compact Disc and digital tape. Stereo was introduced just 25 years ago, but how many monophonic records are found in most individual's record collections? Similarly, digital formats will supersede analog ones, especially since the life span of an analog recording is limited. To replace all consumer playback systems, and record libraries... once again, the question of economics emerges. Sure, some consumers will hold out, but that old record player will not last forever....

To succeed against the proven and entrenched analog technology, digital audio equipment must ultimately be viable on the consumer's bottom line of affordability. Fortunately, digital audio will prove to be a more and more cost-effective technology. Digital technology promises to re-define the relationship of electronic technology and its economics. In the past, price was essentially based on labor and parts count. A professional tape recorder was more expensive because its circuits contained more parts and required more assembly than a consumer version. With digital technology much of the circuitry is contained on integrated circuits. To some degree, more complex circuit designs occupy greater chip areas thus dictating greater chip cost, but overall product price is primarily a function of volume. If a product is mass-produced the manufacturing cost is distributed, and the cost for a single chip can be quite low. Thus, very sophisticated circuits may be integrated and sold cheaply if volume permits.

So-called professional versions of electronic technology might cease to exist because small quantity integrated circuits will not be economically feasible. There is no such thing as a "professional" digital watch because they all use essentially the same chip. Whereas with analog technology, a professional watch might make sense, with digital, a special chip with truly professional features would be prohibitively expensive. With a generic chip the only difference in price between digital watches is a gold case versus a plastic one. The same might be true in audio.

The distinction between consumer and professional equipment might become slight as mass produced consumer products with sophisticated circuitry begin to set the true pace of technical development. Early examples of this were the digital audio processors, devices designed for the consumer market which surprised even their manufacturers by suddenly surpassing the equality of much professional recording gear. With the shifting economics of digital technology, consumer audio equipment will be able to feature the newest technology at an affordable price because of the mass-production of the integrated circuits utilized in the products.

8.2-3 The Decision • The introduction of a wholly new technology will be an interesting spectacle to watch. Many recording studios have committed themselves to digital equipment, but as we have noted, success of the technology will hinge largely on the consumer's perception, and bank account. For all its limitations, analog recording has served us for over 100 years. Perhaps the most significant sign of its utility has been its ability to readily adapt itself to technological advances yet maintain an affordable price; this pays high tribute to Edison's groove.

One transition obstacle will be the availability of software. The first CD catalog issued at the time of the system's introduction in the United States listed about 200 titles; whereas, the LP catalog contained over 40,000 titles. Correspondingly, the entire initial production volume of CDs was far fewer than sales of single popular LP releases. Yet manufacturers have predicted that CD sales will surpass those of LP within 5 years, and that in 10 years the CD will be the dominant medium for home, car, and portable listening. Could such an initially expensive, and in some circles, technically controversial medium so rapidly gain supremacy? That is the question upon which rests millions of dollars of research and development, and revitalization of the recording industry.

What is important to the consumer when the consumer compares digital to analog? Overall, fidelity is perhaps the main concern, but other questions such as permanence, convenience, and of course price, all enter into the decision-making process. For a better understanding of the consumer's perspective, let's consider these criteria. Fidelity is on everyone's mind— everyone from the concert-goer who desires acoustically excellent seats, to the home listener in his or her favorite chair precisely aligned between two loudspeakers. If digital audio recordings sound better, then other things being equal, they will succeed. Does digital yield higher fidelity? That, of course, is a loaded question. Specifications traditionally used to objectively answer the question are perhaps not applicable. On the face of it, digital's specifications are better than analog's. Digital offers an extended and flat frequency response, a very large dynamic range is realizable, distortion is very low, wow and flutter are confined to the minute variations in a crystal oscillator, channel separation is very high, tape print-through and disc pre-echo are nonexistent. With traditional specifications, digital is clearly superior, but those tests were designed to measure the performance of analog devices. Surely additional measurements are needed to fairly appraise digital audio devices and compare them to analog devices.

Considerations other than sonic fidelity sometimes take precedence; the compact cassette eclipsed reel to reel recorders because of convenience; the LP pushed aside the 78 rpm disc because of longer playing time. Another important consideration is permanence. Analog is a notoriously fragile storage technology; analog discs are destroyed by the very pickup used to play them, and analog tapes disintegrate in storage merely from the passing of time. Digital is more durable and enduring; the laser light playback of CDs makes them impervious to the effects of playback. Digital magnetic recordings are robust and could be reclaimed with minimal loss. Duplication can be accomplished with much greater accuracy with digital. Ironically, the quality of digital duplication may ultimately hurt the recording industry as it places master material in the hands of pirates....

Signal processing is similarly facilitated with digital technology. Hardware processors are replaced with software programs thus greater flexibility is obtained and only the creativity of the programmer limits the possibilities. Delay, reverberation, equalization, compression, expansion, phasing, panning, and any heuristic function could be programmed and superimposed on the signal. The performance of the final interface is in the reproduction chain; the electrical/acoustical interface in the consumer's listening room could be vastly improved as automated electroacoustic systems tailor the playback to the room's acoustics. Error detection and correction further demonstrate the utility of the software nature of digital audio technology, in this case to accomplish something not even feasible with analog equipment.

Digital audio equipment will be sold through conversions of analog systems to digital, new sales, as well as entirely new markets which are sure to be revealed. For example, consider the many vast archives of analog recordings slowly disintegrating with age. To ensure their safe preservation, all of this material will have to be re-recorded onto a digital format. Additionally, entirely new forms of software distribution, such as central library pay systems, will flourish. Satellite distribution of digital audio promises to revolutionize music publishing. The possibilities are enormous, digital technology will completely re-define the economics of the music industry.

8.3 State of the Art

Even digital electronics cannot be viewed as the ultimate technology for audio recording. We are already pushing the practical limits for microelectronic fabrication density and gate switching speeds. Many researchers are turning their attention from electronics to photonics, in which light is the information carrier instead of electrons. Fiber optics has shown itself to be one of the first products of this science and has already begun to affect the communications industry. Similarly, laser technology is advancing rapidly. As CD players already demonstrate, applications to the audio industry are forthcoming; everything from fiber optic microphones and guitar pickups, to laser loudspeakers will introduce yet another technology into audio evolution.

At the far end of the spectrum are the exciting advances in computer programming. Computers are achieving meaningful artificial intelligence in which the nature of problem solving evolves from computation to reasoning. It is not unlikely to expect a symbiotic relationship between computers and brains. The perception and enjoyment of music occurs in the brain which suggests direct access through a properly interfaced digital music/thinking device.

Thus, today's most elaborate digital tape recorders will some day deserve only the same charm we now delegate to primitive analog equipment. That is inevitable because any technology is inherently only a forerunner to more sophisticated devices. The idea that everything modern simply represents a transition toward antiquity is a little difficult to comprehend. It is man's great limitation to observe the mythology of state of the art. He believes in the modernness of any new product and that it somehow embodies the last stage of that product's development; its newness makes it hard to envision anything newer. Even if he avoids that delusion, he accepts the myth to the extent that he decides that this product is so new and so advanced that he has to have it, he cannot wait for it to get better, he has to buy it now. Of course, that instinct fuels technological innovation. Whether stimulated by the profit motive or more lofty scientific ideals, there is always someone looking ahead to the next last word in technology.

Our sometimes limited conception of technological evolution has played tricks with our perception of digital audio. When the promise of digitally recorded music was first proposed, it was partly viewed as the last word of audio science, a method so perfect as to preclude any future development. Great debates sprang up concerning sampling rates, quantization resolution, and formats. The emphasis sometimes seemed to be on finalizing a digital scheme, standardizing it, and somehow halting development. Of course, such an endeavor was pointless, just as analog audio technology underwent a continual process of self renewal, digital audio will similarly undergo changes to participate in our evolving expectations of recorded music. The point is that controversy over digital standards is relatively moot, except when it is accepted beforehand that any standard must be a transitory agreement. It is obvious to state that digital will completely supplant analog as our recording technology, and that succeeding generations of digital standards will make obsolete those preceding them. There is no state of the art, at some point in time we each give in, put our money down, and accept our purchased place in history. Just as Edison started something, engineers of digital audio equipment are today starting something. Everything is transitory, ashes to ashes, rust to rust.

8.3-1 Antediluvian Deconvolution • As for now, we are just entering a new era in audio. The opportunity for the development of digital audio technology is as exciting as it must have been in the very first years of analog audio which saw the fast and furious appearance of startling inventions. But because of the accelerating nature of technology, a simple time line through which we can trace the development of technology over the centuries is no longer applicable. The multiplicity of inventions and their interactions negates the possibility of an algorithm to predict overall development. The time it takes to complete any project is undermined by technology's advance. Enrico Caruso completed his great recordings over half a century ago, and now we use deconvolution techniques to remove the unwanted historical artifacts and thus uncover the sound of his voice. With the issuing of these recordings we manufacture history totally out of the context of time, with recordings that never before existed as they do now. Even as we process Caruso's recorded repertoire, the technology improves, and by the time we are finished, we have to start all over again because the early new releases of the dead artist are not as good as his later new releases. Technology yields that kind of opportunity; with technology the recursions are endless. Everything that technology affects has to be re-evaluated as quickly as technology changes our perception of it.

8.3-2 Leibnitz, Edison, Nyquist • The page in Leibnitz' notebook dated March 15, 1679 is even more of a calligrapher's nightmare than the rest. Evidently, the mathematician/philosopher was excited that day, a little too eager to document an idea that had just occurred to him. His quill pen sprayed ink all over the page. He wrote 1 and 1, 2 and 10, 3 and 11, then he screwed up and scratched something out, and wrote 4 and 100. He followed the progression through to 17 then realized that 16 was an important cadence. He crossed it out and started a new line, and wrote the binary code for 17 through 32. Then in the margin, just to make sure, he continued through to 91 and 1011011, then ran out of paper.

The telegraphic repeater was one of young Edison's early inventions; it used paraffin-coated paper to store and repeat high speed telegraphic messages. A signal was recorded by a stylus attached to the telegraphic arm by embossing the Morse code dots and dashes into the paraffin, then replayed. On July 18, 1877, suddenly inspired to make a small experiment, Edison attached a diaphragm in place of the telegraphic arm, ran a strip of paraffin tape under the stylus, shouted into the horn, then ran the strip through again. In his own words, "We heard a distinct sound, which a strong imagination might have translated into the original Halloo!"

At the winter convention of the American Institute of Electrical Engineers, in New York, February 13–17, 1928, a paper entitled "Certain Topics in Telegraph Transmission Theory" was delivered by an engineer from American Telephone and Telegraph Co. Nyquist presented the "...results of theoretical studies of telegraphic systems which have been made from time to time..." and apologized that it was necessary "...to include a certain amount of material which is already well known to telegraphic engineers." At least one of his points was not well known, or at least well recognized. He stated, "...for any given deformation of the received signal the transmitted frequency range must be increased in direct proportion to the signaling speed, and the effect of the system at any corresponding frequencies must be the same. The conclusion is that the frequency band is directly proportional to the speed."

I hope that this book has fulfilled its intended purpose of introducing readers to the technology of digital audio, and illuminating some of the great potential of this young science.

Bibliography

Chapter 1

Backus, J., The Acoustical Foundations of Music, W.W. Norton, 1969.

Davis, D., Davis, C., Sound System Engineering, Howard W Sams & Co., Inc., 1974.

Eargle, J., Sound Recording, Van Nostrand, 1980.

Helmholtz, H., On The Sensations of Tone, Dover, 1954.

Kinsler, L.E., Frey, A.R., Fundamentals of Acoustics, John Wiley and Sons, 1962.

Olson, H., Music, Physics and Engineering, Dover, 1967.

Pohlmann, K.C., "A Binary Beginning," Audio, April, 1984.

Pohlmann, K.C., "Take A Number," Audio, May, 1984.

Pohlmann, K.C., "Number Systems," Mix, September, 1984.

Pohlmann, K.C., "Digital Algebra," Mix, October, 1982.

Strong, W.J., Plitnik, G.R., Music, Speech, High Fidelity, Soundprint, 1983.

Traylor, J.G., Physics of Stereo/Quad Sound, Iowa State University Press, 1977.

Tremaine, H., The Audio Cyclopedia, Howard W. Sams & Co., Inc., 1974.

Woram, J.W., The Recording Studio Handbook, Elar, 1982.

Chapter 2

Blesser, B., "Digitization of Audio," JAES, October, 1978.

Blesser, B., "Elementary and Basic Aspects of Digital Audio," Digital Audio Collected Papers, AES, 1983.

Clarke, A.B., Disney, R.L., Probability and Random Processes for Engineers and Scientists, John Wiley and Sons, 1970.

Feldman, L., "Digital Audio Myths Debunked," Audio's Digital Buying Guide, 1983.

Lagadec, R., "Digital Sampling Frequency Conversion," Digital Audio Collected Papers, AES, 1983.

Nyquist, H., "Certain Topics in Telegraph Transmission Theory," Trans. AIEE, April, 1928.

Oppenheim, A.V., ed., Applications of Digital Signal Processing, Prentice-Hall, 1978.

Oppenheim, A.V., Schafer, R.W., Digital Signal Processing, Prentice-Hall, 1975.

Pohlmann, K.C., "Almost Free Samples," Audio, January, 1984.

Pohlmann, K.C., "Bit By Bit," Audio, September, 1984.

Pohlmann, K.C., "Fold Over Beethoven," Audio, October, 1984.

Pohlmann, K.C., "Sampling and Quantizing," Mix, January, 1983.

Pohlmann, K.C., "A Digital Audio System," Mix, August, 1983.

Rabiner, L.R., "Digital Techniques for Changing the Sampling Rate of a Signal," Digital Audio Collected Papers, AES, 1983.

Shannon, C.E., "A Mathematical Theory of Communication," Bell System Technical Journal, October, 1968.

Stockham, T.G., Jr., "The Promise of Digital Audio," Digital Audio Collected Papers, AES, 1983.

Talambiras, R., "Some Considerations in the Design of Wide-Dynamic-Range Audio Digitization Systems," AES preprint 1226, 1977.

Tanabe, H., Wakuri, T., "On the Quality of Some Digital Audio Equipment Measured by the High Accuracy Dynamic Distortion Measuring System," AES preprint 1909, 1982.

Vanderkooy, J., Lipshitz, S.P., "Resolution Below the Least Significant Bit in Digital Audio Systems with Dither," JAES, March, 1984.

Chapter 3

Blesser, B., "The Digitization of Audio," JAES, October, 1978.

Blesser, B., "Elementary and Basic Aspects of Digital Audio," Digital Audio Collected Papers, AES, 1983.

Blesser, B., "Advanced Analog-to-Digital Conversion and Filtering: Data Conversion," Digital Audio Collected Papers, AES, 1983.

Carr, J.J., Digital Interfacing with An Analog World, Tab, 1978.

Doi, T.T., "Recent Progress in Digital Audio Technology," Digital Audio Collected Papers, AES, 1983.

Hamill, D.C., "Transient Response of Audio Filters," Wireless World, August, 1981.

Hnatek, E.R., A User's Handbook of D/A and A/D Converters, John Wiley and Sons, 1976.

Hoeschele, D., Analog-to-Digital, Digital-to-Analog Conversion Techniques, John Wiley and Sons, 1968.

Jayant, N., Rabiner, L., "The Application of Dither to the Quantization of Speech Signals," Bell System Technical Journal, vol. 51, 1972.

Jones, R., "Time Domain Considerations in Digital Audio Systems," Mix, May, 1984.

Lagadec, R., Stockham, T.G. Jr., "Dispersive Models for A-to-D and D-to-A Conversion Systems," AES preprint 2097, 1984.

Matthews, M., The Technology of Computer Music, MIT press, 1969.

Meyer, J., "Time Correction of Anti-Aliasing Filters Used in Digital Audio Systems," JAES, March, 1984.

Nakajima, H., Doi, T.T., Fukuda, J., Iga, A., Digital Audio Technology, Tab, 1983.

Peled, A., Liu, B., Digital Signal Processing; Theory, Design, and Implementation, John Wiley and Sons, 1976.

Picot, J.P., Introduction a l'Audio Numerique, Editions Frequences, 1984.

Pohlmann, K.C., "A/D and D/A Conversion," Mix, July, 1983.

Rabiner, L.R., Gold, B., Theory and Applications of Digital Signal Processing, Prentice-Hall, 1975.

Salman, W.P., Solotareff, M.S., Le Filtrage Numerique, Eyrolles, 1978.

Stockham, T.G., Jr., "A/D and D/A Converters: Their Effect on Digital Audio Fidelity," AES proc., October, 1971.

Vanderkooy, J., Lipshitz, S.P., "Resolution Below the Least Significant Bit in Digital Audio Systems with Dither," JAES, March, 1984.

Micro Networks Company, 1980 Product Guide and Applications Manual.

Chapter 4

Adams, R.W., "dbx Model 700 Design Parameters and Systems Implementation," Recording Engineer/Producer, October, 1982.

Baldwin, G.L., Tewksbury, S.K., "Linear Delta Modulation Integrated Circuit with 17 MegaHertz/S Sampling Rate," IEEE trans. on Communications, July, 1974.

Blesser, B., "The Digitization of Audio," JAES, October, 1978.

Carr, J.J., Digital Interfacing with an Analog World, Tab, 1978.

Doi, T.T., "Recent Progress in Digital Audio Technology," Digital Audio Collected Papers, AES, 1983.

Elen, R., "dbx Model 700 Digital Audio Processor," Studio Sound, February, 1983.

Gundry, K.J., Robinson, D.P., Todd, C.C., "Recent Developments in Digital Audio Techniques," AES preprint 1956, 1983.

Hnatek, E.R., A User's Handbook of D/A and A/D Converters, John Wiley and Sons, 1976.

Hoeschele, D., Analog-to-Digital, Digital-to-Analog Conversion Techniques, John Wiley and Sons, 1968.

Nakajima, H., Doi, T.T., Fukuda, J., Iga, A., Digital Audio Technology, Tab, 1983.

Oppenheim. A.L., Schafer, R.W., Digital Signal Processing, Prentice-Hall, 1975.

Pohlmann, K.C., "A/D and D/A Conversion," Mix, July, 1983.

Pohlmann, K.C., "Delta Modulation," Mix, September, 1983.

Schindler, H.R., "Delta Modulation," IEEE Spectrum, October, 1970.

Schott, W., "Philips Oversampling System for Compact Disc Decoding," Audio, April, 1984.

Steele, R., Delta Modulation Systems, Halsted Press, 1975.

Stockham, T.G., Jr., "A/D and D/A Converters: Their Effect on Digital Audio Fidelity," AES proc., October, 1971.

Tyler, L.B., "One Year Later: A Progress Report on the dbx Model 700 Digital Audio Processor," Recording Engineer/Producer, October, 1983.

Van de Plassche, R.J., Dijkmans, E.C., "A Monolithic 16 Bit D/A Conversion System for Digital Audio," Digital Audio Collected Papers, AES, 1983.

Micro Networks Company, 1980 Product Guide and Applications Manual.

Chapter 5

Borwick, J., ed., Sound Recording Practice, Oxford University Press, 1977.

Feher, K., Digital Communications: Satellite/Earth Station Engineering, Prentice-Hall, 1981.

Fink, D.G., ed., Color Television Standards, McGraw-Hill, 1955.

Gross, L.S., The New Television Technologies, William C. Brown, 1983.

Ingebretsen, R.B., Stockham, T.G., Jr., "Random-Access Editing of Digital Audio," JAES, March, 1984.

Johnson, C.E., Jr., "The Promise of Perpendicular Magnetic Recording," Byte, March, 1983.

Keene, S., "The Soundstream Digital Music Computer: Recording, Editing and Beyond," dB, January, 1983.

Lagadec, R., "New Concepts in Digital Audio Editing," AES preprint 2096, 1984.

Lagadec, R., "Labels and Their Formatting in Digital Audio Recording and Transmission," AES preprint 2002, 1983.

Lagadec, R., "Proposed Guidelines For an Auxiliary Data Track on Digital Audio Tape," internal Studer memo, March, 1984.

Lagadec, R., Ginsberger, H.P., Brandes, C., "Design of a Professional 2-Channel Stationary Head Digital Audio Recorder," AES preprint 2095, 1984.

Langhans, R., Shumila, M., "Digital Audio Transmission System Using Satellite Distribution," AES preprint 2018, 1983.

Lemke, J.V., "The State of the Art in High-Density Magnetic Recording," Digital Audio Collected Papers, AES, 1983.

Lowman, C., Magnetic Recording, McGraw-Hill, 1972.

Moran, T., "New Developments in Floppy Disks," Byte, March, 1983.

Nakajima, H., Odaka, K., "A Rotary Head High Density Digital Audio Tape Recorder," IEEE trans, on Consumer Electronics, August, 1983.

Nakajima, H., Doi, T.T., Fukuda, J., Iga, A., Digital Audio Technology, Tab, 1983.

Picot, J.P., Introduction a l'Audio Numerique, Editions Frequencies, 1984.

Pohlmann, K.C., "The Futures Market," Audio, August, 1984.

Rothchild, E.S., "Optical Memory: Data Storage By Laser," Byte, October, 1984.

Rothchild, E.S., "Optical-Memory Media," Byte, March, 1983.

Schouhamen-Immink, K.A., Brant, J.J.M., "Experiments Toward an Erasable Compact Disc Digital Audio System," JAES, July/August, 1984.

Shimek, Y., Kato, M., Tsuji, S., Ishida, K., Matsushima, H., "Digital Recording and Reproducing Techniques with Thin Film Head for Digital Audio Tape Recorder," AES preprint 2027, 1983.

Simmons, W., "Analog Mastering Tape Versus Digital Mastering Tape," dB, January, 1983.

Tamaka, K., "Tape Formats and Multi-Track Formats," Digital Audio Collected Papers, AES, 1983.

Tanaka, K., "A Simple Explanation of the Tape Format of the Mitsubishi Digital Audio Recorders," Mitsubishi press release, 1984.

Tremaine, H., The Audio Cyclopedia, Howard W Sams & Co.,Inc., 1974.

Viterbi, A.J., "Coding and Interleaving for Correcting Burst and Random Errors in Recording Media," Digital Audio Collected Papers, AES, 1983.

White, G., Video Recording, Butterworth and Co., 1972.

CADA System, Sony press release, March, 1984.

Digital Audio Stationary Head (DASH) Format, Sony press release, October, 1983.

Direct Read After Write (DRAW) System, Sony press release, March, 1984.

Highly Sensitive Satellite Broadcast Receiver System, Sony press release, 1983.

Magneto-Optical Disc, Sony press release, March, 1984.

Outline of BS-II, Sony press release, March, 1984.
PCM 3102 2 Channel Digital Audio Recorder, Sony press release, March, 1984.

Chapter 6

Berlekamp, E.R., Algebraic Coding Theory, McGraw-Hill, 1968.
Berlekamp, E.R., "Error Correcting Code for Digital Audio," Digital Audio Collected Papers, AES, 1983.
Booth, T.L., Digital Networks and Computer Systems, John Wiley and Sons, 1971.
Doi, T.T., "Error Correction for Digital Audio Recordings," Digital Audio Collected Papers, AES, 1983.
Gallager, R.G., Information Theory and Reliable Communication, Wiley, 1968.
Hagelbarger, O.W., "Recurrent Codes: Easily Mechanized Burst-Correcting, Binary Codes," Bell System Tech. Journal, vol. 38, 1959.
Lin, S., An Introduction to Error-Correcting Codes, Prentice-Hall, 1970.
McEliece, R.J., The Theory of Information and Coding, Addison-Wesley, 1977.
Nakajima, H., Doi, T.T., Fukuda, J., Iga, A., Digital Audio Technology, Tab, 1983.
Osborne, A., An Introduction to Microcomputers—Volume One, Adam Osborne and Associates, 1977.
Peterson, W.W., "Error-Correcting Codes," Scientific American, February, 1962.
Roth, C.H., Jr., Fundamentals of Logic Design, West, 1979.
Stockham, T.G., Jr., "The Promise of Digital Audio," Digital Audio Collected Papers, AES, 1983.
Viterbi, A.J., "Coding and Interleaving for Correcting Burst and Random Errors in Recording Media," Digital Audio Collected Papers, AES, 1983.
Viterbi, A.J., Coherent Communication, McGraw-Hill, 1966.
Vries, L.B., Odaka, K., "CIRC—The Error Correcting Code for the Compact Disc Digital Audio System," Digital Audio Collected Papers, AES, 1983.
Woram, J.M., The Recording Studio Handbook, Elar, 1982.

Chapter 7

Birchall, S.T., "The CD, It's the Pits," Digital Audio Magazine, January, 1985.
Brewer, B., "Let Your CD's do the Walking," Digital Audio Magazine, December, 1984.
Carasso, M.G., Peck, J.B.H., Sinjou, J.P., "The Compact Disc Digital Audio System," Philips Technical Review, vol. 40, no. 6, 1982.
Feldman, "Beware Tall Tales about Compact Discs," Sound Canada, January, 1984.
Goedhart, D., Van de Plassche, R.J., Stikvoort, E.F., "Digital-to-Analog Conversion in Playing a Compact Disc," Philips Technical Review, vol. 40, no. 6, 1982.
Heemskerk, J.P.J., Schouhamer-Immink, K.A., "Compact Disc System Aspects and Modulation," Philips Technical Review, vol. 40, no. 6, 1982.
Matull, J., "IC's for Compact Disc Decoders, Electrical Components and Applications," May, 1982.

Miyaoka, S., "Manufacturing Technology of the Compact Disc," *Digital Audio Collected Papers*, AES, 1983.

Motokane, E., "Compact Disc Player Operation," *Hitachi internal memo*, 1983.

Ogowa, H., Schouhamer-Immink, K.A., "EFM—The Modulation Method for the Compact Disc Digital Audio System," *Digital Audio Collected Papers*, AES, 1983.

Pohlmann, K.C., "Compact Disc Video," *Mix*, September, 1984.

Timmermans, H.H.J., Vries, L.B., "Error Correction and Concealment in the Compact Disc System," *Philips Technical Review*, vol. 40, no. 6, 1982.

Verkaik, W., "Compact Disc (CD) Manufacturing—An Industrial Process," *Digital Audio Collected Papers*, AES, 1983.

Vries, L.B., Odaka, K., "CIRC—The Error Correcting Code for the Compact Disc Digital Audio System," *Digital Audio Collected Papers*, AES, 1983.

Compact Disc ROM Format, Philips and Sony press release, October, 1983.

Compact Disc Digital Filter Design, Philips press release, 1983.

Hitachi Service Manual for DA-1000, March 1983.

Chapter 8

Conot, R.E., *A Streak of Luck*, Seaview, 1979.

Krell, D.F., ed., *Heidegger—Basic Writings*, Harper and Row, 1977.

Nyquist, H., "Certain Aspects in Telegraphic Transmission Theory," *Trans. AIEE*, April, 1928.

Pohlmann, K.C., "An Historical Perspective," *Mix*, October, 1983.

Pohlmann, K.C., "Celebrating a New Issue," *Digital Audio*, September, 1984.

Stockham, T.G., Jr., Cannon, T., Ingebretsen, R., "Blind Deconvolution Through Digital Signal Processing," *proc. IEEE*, April, 1975.

Index

MORE
FROM
SAMS

☐ The Sams Hookup Book: Do-It-Yourself Connections for Your VCR
Howard W. Sams Engineering Staff
Here is all the information needed for simple to complex hook ups of home entertainment equipment. This step-by-step guide provides instructions to hook up a video cassette recorder to a TV, cable converter, satellite receiver, remote control, block converter, or video disk player.
ISBN: 0-672-22248-5, $4.95

☐ Audio IC Op-Amp Applications (3rd Edition) *Walter G. Jung*
This updated version of a classic reference will be welcomed by recording and design engineers and hobbyists using audio signal processing. This new edition covers the changes that have marked the Op-Amp field over the last few years and includes new devices such as the OP-27/37 and application ICs for automobile stereo and audio testing. The update also includes new applications circuitry to illustrate current usage, among them differential input/output IC devices. Jung is a recognized expert in his field and is the author of the definitive *IC Op-Amp Cookbook*.
ISBN: 0-672-22452-6, $17.95

☐ Basics of Audio-Visual Systems Design *Raymond Wadsworth*
Newcomers to the audio-visual industry will find indispensable information presented here. System designers, architects, contractors, equipment suppliers, students, teachers, consultants. . . all will find these NAVA-sanctioned fundamentals pertinent to system design procedures. Topics include image format, screen size and performance, front versus rear projection, projector output, audio, and the effective use of mirrors.
ISBN: 0-672-22038-5, $15.95

☐ CD-I and Interactive Videodisc Technology *Steve Lambert and Jane Sallis, Editors*
This comprehensive reference guide explains how to program interactive videodiscs, CD-I, and digital interactive formats including CD-ROM and optical storage. It also discusses how the videodisc player and computer communicate, the pros and cons of authoring languages, and interface cards. Additional topics include efficiency statistics, peripherals, types of networking systems, and surrogate computer controllers.
ISBN: 0-672-22513-1, $24.95

☐ The Complete Guide to Car Audio
Martin Clifford
Car audio systems are becoming almost as complex as home sound systems. Choices abound, but quality varies. This book describes car audio system components and helps you plan your own system. Sections on installation, noise control, and theft protection complete this valuable reference.
ISBN: 0-672-21820-8, $9.95

☐ John D. Lenk's Troubleshooting & Repair of Audio Equipment *John D. Lenk*
This manual provides the most up-to-date data available and a simplified approach to practical troubleshooting and repair of major audio devices. Coverage includes dual cassette decks, compact disc players, linear-tracking turntables, frequency-synthesized AM/FM tuners, IC amplifiers, and loudspeakers.
ISBN: 0-672-22517-4, $21.95

☐ Modern Recording Techniques (2nd Edition)
Robert E. Runstein and David Miles Huber
Engineers and students alike will find this a valuable guide to state-of-the-art developments and practices in the recording industry. This revised edition reflects all the latest equipment, controls, acoustics, and digital effect devices being used in modern recording studios. It explores the marriage of video and audio multi-track studios and illustrates sound and studio capabilities and limitations.
ISBN: 0-672-22451-8, $18.95

☐ Principles of Digital Audio
Ken C. Pohlmann
Here's the one source that covers the entire spectrum of audio technology. Includes the compact disk, how it works, and how data is encoded on it. Illustrates how digital audio improves recording fidelity. Starting with the fundamentals of numbers, sampling, and quantizing, you'll get a look at a complete audio digitization system and its components. Gives a concise overview of storage mediums, digital data processing, digital/audio conversion, and output filtering. Filled with diagrams and formulas, this book explains digital audio thoroughly, yet in an easy-to-understand style.
ISBN: 0-672-22388-0, $19.95

MORE
FROM
SAMS

☐ Handbook for Sound Engineers: The New Audio Cyclopedia
Glen Ballou, Editor-in-Chief
Special attention is paid to developments in audio electronics, circuits, and equipment. There is also an in-depth examination of disk, magnetic, and digital recording and playback.
ISBN: 0-672-21983-2, $79.95

☐ Sound System Engineering (2nd Edition)
Don Davis and Carolyn Davis
This reference guide is written for the professional audio engineer. Everything from audio systems and loudspeaker directivity to sample design applications and specifications is covered in detail.
ISBN: 0-672-21857-7, $39.95

☐ Understanding IC Operational Amplifiers (3rd Edition)   *Roger Melen and Harry Garland*
Technological advances are bringing us ever closer to the ideal op amp. This book describes that ideal op amp and takes up monolithic to integrated circuit op amp design. Linear and nonlinear applications are discussed, as are CMOS, BIMOS, and BIFET op amps.
ISBN: 0-672-22484-4, $12.95

☐ How to Build Speaker Enclosures
Alexis Badmaieff and Don Davis
A practical guide to the whys and hows of constructing high quality, top performance speaker enclosures. A wooden box alone is not a speaker enclosure — size, baffling, sound insulation, speaker characteristics, and crossover points must all be carefully considered.
ISBN: 0-672-20520-3, $6.95

☐ Reference Data for Engineers: Radio, Electronics, Computer, and Communications (7th Edition)
Edward C. Jordan, Editor-in-Chief
Previously a limited private edition, now an internationally accepted handbook for engineers. Includes over 1300 pages of data compiled by more than 70 engineers, scientists, educators and other eminent specialists in a wide range of disciplines. Presents information essential to engineers, covering such topics as: digital, analog, and optical communications; lasers; logic design; computer organization and programming, and computer communications networks. An indispensable reference tool for all technical professionals.
ISBN: 0-672-21563-2, $69.95

☐ Introduction to Professional Recording Techniques
Bruce Bartlett, The John Woram Audio Series
This all-inclusive introduction to the equipment and techniques for state-of-the-art recording—whether in residences or professional studios or on location— offers a wealth of valuable information on topics not found in other books on audio recording. Geared primarily for the audio hobbyist or aspiring professional, this book delivers a comprehensive discussion of recording engineering and production techniques, including special coverage of microphones and microphone techniques, sampling, sequencing, and MIDI. It provides up-to-date coverage of monitoring, special effects, hum prevention, and spoken-word recording, as well as special sections on recognizing good sound and troubleshooting bad sound.
ISBN: 0-672-22574-3, $18.95

☐ Home Video Handbook (3rd Edition)
Charles Bensinger
Video system integration and use are the themes here. Learn how to hook the components together to make your camera, VCR, videodisc, projection TV, and/or satellite receiver a system you can be proud of. Includes tips on how to buy the best equipment for your needs and receive the greatest benefits.
ISBN: 0-672-22052-0, $13.95

☐ VCR Troubleshooting & Repair Guide
Robert C. Brenner
With approximately 25 million video cassette recorders on the market, this long-awaited book will help owners repair these popular machines when they break down. This helpful troubleshooting guide is for the electronics hobbyist, layperson, or technician who needs a preventive maintenance and troubleshooting reference for VCRs. Limited electronics experience is required to use it, but more sophisticated service and repair functions are discussed, and valuable information for the service technician is included.
ISBN: 0-672-22507-7, $19.95

☐ Video Cameras: Theory & Servicing
Gerald P. McGinty
This entry-level technical primer on video camera servicing gives a clear, well-illustrated presentation of practical theory. From the image tube through the electronics to final interface, all concepts are fully discussed.
ISBN: 0-672-22382-1, $18.95

MORE
FROM
SAMS

☐ **Video Production Guide**   *Lon McQuillin*
For those who want to learn how video production really works, this book contains real-world applications. Pre-production planning, creativity and organization, people handling, single and multi-camera studio and on location production, direction techniques, editing, special effects, and distribution of the finished production are addressed. This book is designed for working and aspiring producers/directors, broadcasters, schools, CATV personnel, and others in the industry.
ISBN: 0-672-22053-9, $28.95

☐ **Video Tape Recorders (2nd Edition)**
Harry Kybett
This book shows you how to operate and service helical VTRs and includes numerous examples of recorder circuitry and mechanical transport systems.
ISBN: 0-672-21521-7, $14.95

☐ **Video Scrambling and Descrambling for Satellite and Cable TV**
Rudolf F. Graf and William Sheets
Learn the secrets of signal scrambling and descrambling (encoding and decoding) from the experts. The book discusses the theory and techniques needed to understand how over-the-air and cable signals are decoded and encoded. Projects are

included such as building a scrambler and descrambler and how to build a video test generator with scrambling capability.
ISBN: 0-672-22499-2, $19.95

☐ **Troubleshooting with the Oscilloscope (4th Edition)**   *Robert G. Middleton*
One of the quickest and least costly ways to troubleshooting most electronic equipment is to use an oscilloscope — properly. In this book, now in its fourth edition, the author not only provides correct step-by-step procedures on the use of an oscilloscope but combines these with specific facts of television receiver troubleshooting.
ISBN: 0-672-21738-4, $11.95

☐ **Know Your Oscilloscope (4th Edition)**
Robert G. Middleton
The oscilloscope remains the principal diagnostic and repair tool for electronic technicians. This book provides practical data on the oscilloscope and its use in TV and radio alignment, frequency and phase measurements, amplifier testing and signal tracing, and digital equipment servicing. Additional material is provided on oscilloscope circuits and accessories. A vital reference for your workbench.
ISBN: 0-672-21742-2, $11.95

Availability, prices, and page counts are subject to change without notice.

Look for these Sams Books at your local bookstore.

To order direct, call 800-428-SAMS or fill out the form below.

- -

Please send me the books whose titles and numbers I have listed below.

_____

_____

_____

_____

_____

Enclosed is a check or money order for $ _____
Include $2.50 postage and handling.
AR, CA, FL, IN, NC, NY, OH, TN, WV residents add local sales tax.

Charge my: ☐ VISA ☐ MC ☐ AE
Account No. _____ Expiration Date _____

Name *(please print)* _____

Address _____

City _____

State/Zip _____

Signature _____
(required for credit card purchases)

Mail to: Howard W. Sams & Co.
Dept. DM
4300 West 62nd Street
Indianapolis, IN 46268

DC 128

SAMS ™